Lecture Notes in Computer Science 9463

More information about this series at http://www.springer.com/series/7407

Mihai Codescu · Răzvan Diaconescu
Ionuţ Ţuţu (Eds.)

Recent Trends in Algebraic Development Techniques

22nd International Workshop, WADT 2014
Sinaia, Romania, September 4–7, 2014
Revised Selected Papers

Springer

Editors
Mihai Codescu
Otto-von-Guericke-Universität Magdeburg
Magdeburg
Germany

Ionuţ Ţuţu
Royal Holloway University of London
Egham
UK

Răzvan Diaconescu
Simion Stoilow Institute of Mathematics
of the Romanian Academy
Bucharest
Romania

ISSN 0302-9743 ISSN 1611-3349 (electronic)
Lecture Notes in Computer Science
ISBN 978-3-319-28113-1 ISBN 978-3-319-28114-8 (eBook)
DOI 10.1007/978-3-319-28114-8

Library of Congress Control Number: 2015957798

LNCS Sublibrary: SL1 – Theoretical Computer Science and General Issues

Printed on acid-free paper

This Springer imprint is published by SpringerNature
The registered company is Springer International Publishing AG Switzerland

In memoriam Joseph A. Goguen

Preface

This volume contains one invited paper and eight selected papers from the 22nd International Workshop on Algebraic Development Techniques (WADT 2014), which took place in Sinaia, Romania, during September 4–7, 2014. The event was dedicated to the memory of Joseph A. Goguen – one of the founding members of the ADT community – who visited Sinaia many times with great pleasure in order to meet with his former student and close friend Răzvan Diaconescu, the local chair of WADT 2014. Sinaia is a beautiful mountain resort that has become a traditional Romanian venue for mathematical and theoretical computer science events, from the second edition of the International Mathematical Olympiad, which was held in 1960, to the more recent Sinaia School on Formal Verification of Software Systems, the Romanian–Japanese Algebraic Specification Workshops, and the Summer School on Language Frameworks, which discussed some of the latest developments related to algebraic specification and in particular to the CafeOBJ language and the K semantic framework.

The algebraic approach to system specification encompasses many aspects of the formal design of software systems. Originally born as formal method for reasoning about abstract data types, it now covers new specification frameworks and programming paradigms (such as object-oriented, aspect-oriented, agent-oriented, logic, and higher-order functional programming) as well as a wide range of application areas (including information systems, concurrent, distributed, and mobile systems). The workshop provided an opportunity to present recent and ongoing work, to meet colleagues, and to discuss new ideas and future trends. Typical topics of interest are:

- Foundations of algebraic specification
- Other approaches to formal specification, including process calculi and models of concurrent, distributed and mobile computing
- Specification languages, methods, and environments
- Semantics of conceptual modeling methods and techniques
- Model-driven development
- Graph transformations, term rewriting, and proof systems
- Integration of formal specification techniques
- Formal testing, quality assurance, validation, and verification

As 22 occurrences of the ADT Workshop can be considered as something noteworthy, a short look back may be allowed. The first workshop took place in 1982 in Sorpesee, followed by Passau (1983), Bremen (1984), Braunschweig (1986), Gullane (1987), Berlin (1988), Wusterhausen (1990), Dourdan (1991), Caldes de Malavella (1992), S. Margherita (1994), Oslo (1995), Tarquinia (1997), Lisbon (1998), Chateau de Bonas (1999), Genoa (2001), Frauenchiemsee (2002), Barcelona (2004), La Roche en Ardenne (2006), Pisa (2008), Etelsen (2010), and Salamanca (2012). With only a few exceptions at the beginning, it also became a tradition to publish selected papers after each workshop in dedicated volumes of Springer's *Lecture Notes in Computer*

Science (LNCS 332, 534, 655, 785, 906, 1130, 1376, 1589, 1827, 2267, 2755, 3423, 4409, 5486, 7137, and 7841) under the title *Recent Trends in Algebraic Development Techniques*. This speaks to the stability of the ADT community and the continuity of the topics of interest. One should realize, however, that some significant transformations took place from 1982 to today. While ADT stood initially for *Abstract Data Types*, it is now (since 1997) the acronym for *Algebraic Development Techniques*, and the list of topics has broadened accordingly in an amazing way.

The scientific program of WADT 2014 consisted of two invited talks by K. Rustan M. Leino (from Microsoft Research, USA) and Christoph Benzmüller (from Freie Universität Berlin, Germany), and of 32 presentations based on selected abstracts of ongoing research. The abstracts were compiled in a technical report made available during the meeting and later published in the Preprint Series of the Institute of Mathematics of the Romanian Academy. As with previous ADT workshops, the authors were invited to submit full papers, which underwent a thorough review process that was managed using EasyChair. Each paper was reviewed by at least two referees. We would like to thank all the authors who submitted papers and who generally showed interest in the subject, as well as the members of the Program Committee and the external referees appointed by them for their work in completing the reviewing process.

The workshop took place under the auspices of IFIP WG1.3, and was organized by former lecturers and students at the Postgraduate Academic Studies School *Şcoala Normală Superioară Bucureşti* (SNSB). We gratefully acknowledge the support offered by IFIP TC1 and the Simion Stoilow Institute of Mathematics of the Romanian Academy (IMAR).

June 2015

Mihai Codescu
Răzvan Diaconescu
Ionuţ Ţuţu

Organization

Steering Committee

Michel Bidoit	CNRS and ENS de Cachan, France
Andrea Corradini	Università di Pisa, Italy
José Luiz Fiadeiro	Royal Holloway University of London, UK
Rolf Hennicker	Ludwig-Maximilians-Universität München, Germany
Hans-Jörg Kreowski	Universität Bremen, Germany
Till Mossakowski (Chair)	Otto-von-Guericke-Universität Magdeburg, Germany
Fernando Orejas	Universitat Politécnica de Catalunya, Spain
Francesco Parisi-Presicce	Università di Roma, Italy
Grigore Roşu	University of Illinois at Urbana-Champaign, USA
Andrzej Tarlecki	University of Warsaw, Poland

Organizing Committee

Mihai Codescu	Otto-von-Guericke-Universität Magdeburg, Germany
Răzvan Diaconescu (Chair)	Simion Stoilow Institute of Mathematics of the Romanian Academy, Romania
Ionuţ Ţuţu	Royal Holloway University of London, UK

Program Committee

The Program Committee of WADT 2014 comprises all the members of the Steering Committee and two of the members of the Organizing Committee: Mihai Codescu, from Otto-von-Guericke-Universität Magdeburg, and Ionuţ Ţuţu, from Royal Holloway University of London.

Additional Reviewers

Bocchi, Laura
Caleiro, Carlos
Goncharov, Sergey
Günther, Stephan
Madeira, Alexandre
Martins, Manuel António
Moore, Brandon
Şerbănuţă, Traian Florin
Ştefănescu, Andrei
Tronci, Enrico
Tuosto, Emilio

Contents

Invited Talk

On Logic Embeddings and Gödel's God

Christoph Benzmüller[1(✉)] and Bruno Woltzenlogel Paleo[2]

[1] Freie Universität Berlin, Berlin, Germany
c.benzmueller@fu-berlin.de
[2] Vienna Technical University, Vienna, Austria
bruno@logic.at

Abstract. We have applied an elegant and flexible logic embedding approach to verify and automate a prominent philosophical argument: the ontological argument for the existence of God. In our ongoing computer-assisted study, higher-order automated reasoning tools have made some interesting observations, some of which were previously unknown.

Logic embeddings provide an elegant means to formalize sophisticated non-classical logics in classical higher-order logic (HOL, Church's simple type theory [14]). In previous work (cf. [4] and the references therein) the embeddings approach has been successfully applied to automate object-level and meta-level reasoning for a range of logics and logic combinations with off-the-shelf HOL theorem provers. This also includes quantified modal logics (QML) [9] and quantified conditional logics (QCL) [3]. For many of the embedded logics few or no automated theorem provers did exist before. HOL is exploited in this approach to encode the semantics of the logics to be embedded, for example, Kripke semantics for QMLs [15] or selection function semantics for QCLs [26].

The embeddings approach is related to labelled deductive systems [18], which employ meta-level (world-)labeling techniques for the modeling and implementation of non-classical proof systems. In our embeddings approach such labels are instead encoded in the HOL logic.

The embedding approach is flexible, because various modal logics (even with multiple modalities or a mix of varying/cumulative domain quantifiers) can be easily supported by stating their characteristic axioms. Moreover, it is relatively simple to implement, because it does not require any modification in the source code of the higher-order prover. A minimal encoding of second-order modal logic KB in TPTP THF syntax [27] — this syntax is accepted by a range of HOL automated theorem provers (ATPs) — is exemplarily provided in Fig. 1.[1]

C. Benzmüller—This work has been supported by the German Research Foundation DFG under grants BE2501/9-1,2 and BE2501/11-1.

[1] Some Notes on THF, which is a concrete syntax for HOL: `$i` and `$o` represent the HOL base types i and o (Booleans). `$i>$o` encodes a function (predicate) type. Predicate application as in $A(X, W)$ is encoded as `((A@X)@W)` or simply as `(A@X@W)`, i.e., function/predicate application is represented by `@`; universal quantification and λ-abstraction as in $\lambda A_{i \to o} \forall W_i (A\, W)$ and are represented as in `^[X:$i>$o]:![W:$i]:(A@W)`; comments begin with `%`.

M. Codescu et al. (Eds.): WADT 2014, LNCS 9463, pp. 3–6, 2015.
DOI: 10.1007/978-3-319-28114-8_1

```
1  %----The base type $i (already built-in) stands here for worlds and
2  %----mu for individuals; $o (also built-in) is the type of Booleans
3  thf(mu_type,type,(mu:$tType)).
4  %----Reserved constant r for accessibility relation
5  thf(r,type,(r:$i>$i>$o)).
6  %----Modal operators not, or, box
7  thf(mnot_type,type,(mnot:($i>$o)>$i>$o)).
8  thf(mnot,definition,(mnot = (^[A:$i>$o,W:$i]:~(A@W)))).
9  thf(mor_type,type,(mor:($i>$o)>($i>$o)>$i>$o)).
10 thf(mor,definition,(mor = (^[A:$i>$o,Psi:$i>$o,W:$i]:((A@W) | (Psi@W))))).
11 thf(mbox_type,type,(mbox:($i>$i>$o)>($i>$o)>$i>$o)).
12 thf(mbox,definition,(mbox = (^[A:$i>$o,W:$i]:![V:$i]:(~(r@W@V) | (A@V))))).
13 %----Quantifier (constant domains) for individuals and propositions
14 thf(mall_ind_type,type,(mall_ind:(mu>$i>$o)>$i>$o)).
15 thf(mall_ind,definition,(mall_ind = (^[A:mu>$i>$o,W:$i]:![X:mu]:(A@X@W)))).
16 thf(mall_indset_type,type,(mall_indset:((mu>$i>$o)>$i>$o)>$i>$o)).
17 thf(mall_indset,definition,(
18     mall_indset = (^[A:(mu>$i>$o)>$i>$o,W:$i]:![X:mu>$i>$o]:(A@X@W)))).
19 %----Definition of validity (grounding of lifted modal formulas)
20 thf(v_type,type,(v:($i>$o)>$o)).
21 thf(mvalid,definition,(v = (^[A:$i>$o]:![W:$i]:(A@W)))).
22 %----Properties of accessibility relations: symmetry
23 thf(msymmetric_type,type,(msymmetric:($i>$i>$o)>$o)).
24 thf(msymmetric,definition,(
25     msymmetric = (^[R:$i>$i>$o]:![S:$i,T:$i]:((R@S@T)=>(R@T@S))))).
26 %----Here we work with logic KB, i.e., we postulate symmetry for r
27 thf(sym,axiom,(msymmetric@r)).
```

Fig. 1. HOL encoding of second-order modal logic KB in THF syntax. Modal formulas are mapped to HOL predicates (with type `$i>$o`); type `$i` now stands for possible worlds. The modal connectives $\neg$ (`mnot`), $\vee$ (`mor`) and $\Box$ (`mbox`), universal quantification for individuals (`mall_ind`) and for sets of individuals (`mall_indset`) are introduced in lines 7–18. Validity of lifted modal formulas is defined in the standard way (lines 20–21). Symmetry of accessibility relation r is postulated in lines 23–26. Hence, second-order KB is realized here; for logic K the symmetry axiom can be dropped.

The given set of axioms turns any TPTP THF compliant HOL-ATP in a reasoning tool for second-order modal logic. A Henkin-style semantics is thereby assumed for both logics: HOL and second-order modal logic.

In recent work [5,6,8] we have applied the embedding approach to verify and automate a philosophical argument that has fascinated philosophers and theologians for about 1000 years: the ontological argument for the existence of God [25]. We have thereby concentrated on Gödel's [19] modern version of this argument and on Scott's [24] modification, which employ a second-order modal logic (S5) for which, until now, no theorem provers were available. In our computer-assisted study of the argument, the HOL provers LEO-II [10], Satallax [13] and Nitpick [12] have made some interesting observations, some of which were unknown so far. This is a landmark result, with media repercussion in a global scale, and yet it is only a glimpse of what can be achieved by combining computer science, philosophy and theology.

We briefly summarize some of these observations: Nitpick confirms that Scott's axioms are consistent, while LEO-II and Satallax demonstrate that Gödel's original, slightly different axioms are inconsistent. As far as we are aware, this is a new result. As experiments with LEO-II revealed, the problem lies in a subtle difference in the definitions of the predicate *essence* (characterizinghe essential properties of an entity) between Gödel and Scott. In recent papers on the ontological argument (see e.g. below), some authors speak of an oversight/flaw by Gödel, some silently replace Gödel's definition without commenting and some simply stay with it. Moreover, instead of using modal logic S5, LEO-II and Satallax can prove the final theorem (that is, $\Box\exists x.G(x)$, necessarily there exists God) already for modal logic KB. This is highly relevant since some philosophers have criticized Gödel's argument for the use of logic S5. Axiom B (symmetry), however, cannot be dropped, which in turn is confirmed by Nitpick. LEO-II and Satallax can also show that Gödel's and Scott's axioms imply what is called the modal collapse: $\phi \supset \Box\phi$. This expresses that contingent truth implies necessary truth (which can even be interpreted as an argument against free will; cf. [25]) and is probably the most fundamental criticism put forward against Gödel's and Scott's versions of the argument. Other theorems that can be shown by LEO-II and Satallax include flawlessness of God and monotheism.

Ongoing and future work concentrates on the systematic study of Gödel's and Scott's proofs. We have also begun to study more recent variants of the argument [1,2,11,16,17,20,21], which claim to remedy some fundamental problems of Gödel's and Scott's proofs, especially the modal collapse [7]. One interesting and very encouraging observation from these studies is, that the argumentation granularity typical of these philosophy papers is already within reach of the capabilities of our higher-order automated theorem provers. This provides good evidence for the potential relevance of the embedding approach (not only) w.r.t. other similar applications in metaphysics.

The long-term goal is to methodically determine the range of logical parameters (e.g., constant vs. varying domains, rigid vs. non-rigid terms, logics KB vs. S4 vs. S5, etc.) under which the proposed variants of the modern ontological argument hold or fail.

There have been few related works [22,23], and they have focused solely on the comparably simpler, original ontological argument by Anselm of Canterbury. These works do not achieve the close correspondence between the original formulations and the formal encodings that can be found in our approach and they also do not reach the same degree of proof automation.

Our work attests the maturity of contemporary interactive and automated deduction tools for HOL and demonstrates the elegance and practical relevance of the embeddings-based approach. Most importantly, our work opens new perspectives towards computational metaphysics.

References

1. Anderson, C.A.: Some emendations of Gödel's ontological proof. Faith Philos. **7**(3), 291–303 (1990)
2. Anderson, C.A., Gettings, M.: Gödel ontological proof revisited. In: Gödel'96: Logical Foundations of Mathematics, Computer Science, and Physics: Lecture Notes in Logic 6, pages 167–172. Springer, (1996)
3. Benzmüller, C.: Automating quantified conditional logics in HOL. In: Proceedings of IJCAI 2013, pp. 746–753, Beijing, China (2013)

4. Benzmüller, C.: A top-down approach to combining logics. In: Proceedings of ICAART 2013, pp. 346–351. SciTePress Digital Library, Barcelona, Spain (2013)
5. Benzmüller, C., Paleo, B.W.: Gödel's God in Isabelle/HOL. Arch. Formal Proofs (2013)
6. Benzmüller, C., Paleo, B.W.: Automating Gödel's ontological proof of God's existence with higher-order automated theorem provers. In: ECAI 2014. Frontiers in AI and Applications, vol. 263, pp. 163–168. IOS Press (2014)
7. Benzmüller, C., Weber, L., Paleo, B.W.: Computer-assisted analysis of the Anderson-Hájek ontological controversy. Handbook of the 1st World Congress on Logic and Religion, João Pessoa, Brazil (2015)
8. Benzmüller, C., Woltzenlogel Paleo, B.: Interacting with modal logics in the coq proof assistant. In: Beklemishev, L.D. (ed.) CSR 2015. LNCS, vol. 9139, pp. 398–411. Springer, Heidelberg (2015)
9. Benzmüller, C., Paulson, L.: Quantified multimodal logics in simple type theory. Logica Univers. (Spec. Issue Multimodal Logics) **7**(1), 7–20 (2013)
10. Benzmüller, C.E., Paulson, L.C., Theiss, F., Fietzke, A.: LEO-II - a cooperative automatic theorem prover for classical higher-order logic (system description). In: Armando, A., Baumgartner, P., Dowek, G. (eds.) IJCAR 2008. LNCS (LNAI), vol. 5195, pp. 162–170. Springer, Heidelberg (2008)
11. Bjørdal, F.: Understanding Gödel's ontological argument. In: The Logica Yearbook 1998. Filosofia (1999)
12. Blanchette, J.C., Nipkow, T.: Nitpick: a counterexample generator for higher-order logic based on a relational model finder. In: Kaufmann, M., Paulson, L.C. (eds.) ITP 2010. LNCS, vol. 6172, pp. 131–146. Springer, Heidelberg (2010)
13. Brown, C.E.: Satallax: an automatic higher-order prover. In: Gramlich, B., Miller, D., Sattler, U. (eds.) IJCAR 2012. LNCS, vol. 7364, pp. 111–117. Springer, Heidelberg (2012)
14. Church, A.: A formulation of the simple theory of types. J. Symbolic Logic **5**, 56–68 (1940)
15. Fitting, M., Mendelsohn, R.L.: First-Order Modal Logic. Synthese Library, vol. 277. Kluwer, Dordrecht (1998)
16. Fitting, M.: Types, Tableaus, and Gödel's God. Kluwer, Norwell (2002)
17. Fuhrmann, A.: Existenz und Notwendigkeit – Kurt Gödels axiomatische Theologie. In: Logik in der Philosophie, Heidelberg (Synchron) (2005)
18. Gabbay, D.M.: Labelled Deductive Systems. Clarendon Press, Oxford (1996)
19. Gödel, K.: Appx.A: notes in Kurt Gödel's hand. In: [25], pp. 144–145 (2004)
20. Hajek, P.: A new small emendation of Gödel's ontological proof. Stud. Logica: Int. J. Symbolic Logic **71**(2), 149–164 (2002)
21. Hajek, P.: Ontological proofs of existence and non-existence. Stud. Logica: Int. J. Symbolic Logic **90**(2), 257–262 (2008)
22. Oppenheimera, P.E., Zalta, E.N.: A computationally-discovered simplification of the ontological argument. Australas. J. Philos. **89**(2), 333–349 (2011)
23. Rushby, J.: The ontological argument in PVS. In: Proceedings of CAV Workshop "Fun With Formal Methods", St. Petersburg, Russia (2013)
24. Scott, D.: Appx.B: notes in Dana Scott's hand. In: [25], pp. 145–146 (2004)
25. Sobel, J.H.: Logic and Theism: Arguments for and Against Beliefs in God. Cambridge University Press, Cambridge (2004)
26. Stalnaker, R.C.: A theory of conditionals. In: Rescher, N. (ed.) Studies in Logical Theory, pp. 98–112. Blackwell, Oxford (1968)
27. Sutcliffe, G., Benzmüller, C.: Automated reasoning in higher-order logic using the TPTP THF infrastructure. J. Formalized Reasoning **3**(1), 1–27 (2010)

Contributed Papers

An Institutional Foundation for the $\mathbb{K}$ Semantic Framework

Claudia Elena Chiriţă[1](✉) and Traian Florin Şerbănuţă[2]

[1] Department of Computer Science, Royal Holloway University of London, Egham, UK
claudia.elena.chirita@gmail.com
[2] Faculty of Mathematics and Computer Science, University of Bucharest, Bucharest, Romania
traian.serbanuta@fmi.unibuc.ro

Abstract. We advance an institutional formalisation of the logical systems that underlie the $\mathbb{K}$ semantic framework and are used to capture both structural properties of program configurations through pattern matching, and changes of configurations through reachability rules. By defining encodings of matching and reachability logic into the institution of first-order logic, we set the foundation for integrating $\mathbb{K}$ into logic graphs of heterogeneous institution-based specification languages such as HetCasl. This will further enable the use of the $\mathbb{K}$ tool with other existing formal specification and verification tools associated with Hets.

1 Introduction

The $\mathbb{K}$ framework [19] is an executable semantic framework based on rewriting and used for defining programming languages, computational calculi, type systems and formal-analysis tools. It was developed as an alternative to the existing operational-semantics frameworks and over the years has been employed to define actual programming languages, to study runtime verification methods and to develop analysis tools such as type checkers, type inferencers, model checkers and verifiers based on Hoare-style assertions. A comprehensive overview of the framework can be found in [20]. Its associated tool [6] enables the development of modular and executable definitions of languages, and moreover, it allows the user to test programs and to explore their behaviour in an exhaustive manner, facilitating in this way the design of new languages. Driven by recent developments on the theoretical foundations of the $\mathbb{K}$ semantic framework [18,21] and on the established connections with other semantic frameworks and formal systems such as reduction semantics, Hoare logic and separation logic, we propose an institutional formalisation [10] of the logical systems on which the $\mathbb{K}$ framework is based: matching and reachability logic. This would allow us to extend the usage of $\mathbb{K}$ by focusing on its potential as a formal specification language, and furthermore, through its underlying logics, to establish rigorous mathematical relationships between $\mathbb{K}$ and other similar languages, enabling the integration of their verification tools and techniques.

M. Codescu et al. (Eds.): WADT 2014, LNCS 9463, pp. 9–29, 2015.
DOI: 10.1007/978-3-319-28114-8_2

Matching logic [18] is a formal system used to express properties about the structure of mathematical objects and language constructs, and to reason about them by means of pattern matching. Its sentences, called patterns, are built in an inductive manner, similar to the terms of first-order logic, using operation symbols provided by a many-sorted signature, as well as Boolean connectives and quantifiers. The semantics is defined in terms of multialgebras, which interpret patterns as subsets of their carriers. This leads to a ternary satisfaction relation between patterns, multialgebras and elements (or states) of multialgebras.

Unlike first-order logic, matching logic is difficult to formalise faithfully as an institution due to the ternary nature of its satisfaction relation and to the fact that patterns are classified by sorts, much in the way the sentences of branching temporal logics are classified into state or path sentences and evaluated accordingly. We overcome these limitations by relying on the concept of stratified institution developed in [2], which extends institutions with an abstract notion of model state and defines a parameterised satisfaction relation that takes into account the states of models. We further develop this concept by adding classes, which are determined by signatures, associated with sentences, and parameterise both the stratification of models and the satisfaction relation. We show that both matching and computation-tree logic can be described as stratified institutions with classes, and we adapt the canonical construction of an ordinary institution from a stratified one presented in [2] to take into consideration the role of classes.

The main advantage of using stratified institutions with classes to formalise matching logic is that we can extend the construction of reachability logic described in [21] from matching to other logical systems. Reachability logic is a formalism for program verification through which transition systems that correspond to the operational semantics of programming languages can be described using reachability rules; these rules rely on patterns and generalise Hoare triples in order to specify transitions between program configurations (similarly to term-rewrite rules). Therefore, reachability logic can be seen as a language-independent alternative to the axiomatic semantics and proof systems particular to each language. We define an abstract institution of reachability logic over an arbitrary stratified institution with classes such that by instantiating this parameter with matching logic we recover the original notion of reachability.

This paper is based on the Master's thesis of the first author [5], which additionally contains detailed proofs of the results presented herein.

2 Preliminaries

2.1 Institution Theory

The concept of institution [10] formalises the intuitive notion of logic by abstracting the alphabet, syntax, semantics and satisfaction relation. In the following, we assume familiarity with the basics of category theory. The reader is referred to the book [14] of Mac Lane and Eilenberg for further reading.

Definition 1. *An institution* $\mathbf{I} = (\mathbb{S}\mathrm{ig}^{\mathbf{I}}, \mathrm{Sen}^{\mathbf{I}}, \mathrm{Mod}^{\mathbf{I}}, \models^{\mathbf{I}})$ *consists of*

– *a category* $\mathbb{S}\mathrm{ig}^{\mathbf{I}}$ *whose objects are called* signatures*,*

- *a* sentence functor $\mathrm{Sen}^{\mathbf{I}}\colon \mathbb{S}\mathrm{ig}^{\mathbf{I}} \to \mathbb{S}\mathrm{et}$ *giving for every signature* Σ *the set* $\mathrm{Sen}^{\mathbf{I}}(\Sigma)$ *of* Σ-sentences *and for every signature morphism* $\phi\colon \Sigma \to \Sigma'$ *the* sentence translation map $\mathrm{Sen}^{\mathbf{I}}(\phi)\colon \mathrm{Sen}^{\mathbf{I}}(\Sigma) \to \mathrm{Sen}^{\mathbf{I}}(\Sigma')$,
- *a* model functor $\mathrm{Mod}^{\mathbf{I}}\colon (\mathbb{S}\mathrm{ig}^{\mathbf{I}})^{\mathrm{op}} \to \mathbb{C}\mathrm{at}$ *defining for every signature* Σ *the category* $\mathrm{Mod}^{\mathbf{I}}(\Sigma)$ *of* Σ-models *and* Σ-model homomorphisms, *and for every signature morphism* ϕ *the* reduct functor $\mathrm{Mod}^{\mathbf{I}}(\phi)\colon \mathrm{Mod}^{\mathbf{I}}(\Sigma') \to \mathrm{Mod}^{\mathbf{I}}(\Sigma)$,
- *a* satisfaction relation $\models^{\mathbf{I}}_{\Sigma} \subseteq |\mathrm{Mod}^{\mathbf{I}}(\Sigma)| \times \mathrm{Sen}^{\mathbf{I}}(\Sigma)$ *for every signature* Σ,

such that the satisfaction condition

$$M' \models^{\mathbf{I}}_{\Sigma'} \mathrm{Sen}^{\mathbf{I}}(\phi)(\rho) \quad \textit{iff} \quad \mathrm{Mod}^{\mathbf{I}}(\phi)(M') \models^{\mathbf{I}}_{\Sigma} \rho$$

holds for any signature morphism $\phi\colon \Sigma \to \Sigma'$, Σ'*-model* M' *and* Σ*-sentence* ρ.

We may omit the sub/superscripts in the notations of institutions when there is no risk of confusion: for example, $\models^{\mathbf{I}}_{\Sigma}$ may be denoted by $\models$ if the institution $\mathbf{I}$ and the signature Σ are clear. The sentence translation $\mathrm{Sen}^{\mathbf{I}}(\phi)$ and the reduct functor $\mathrm{Mod}^{\mathbf{I}}(\phi)$ may also be denoted by $\phi(\_)$ and $\_\!\upharpoonright_{\phi}$. When $M = M'\!\upharpoonright_{\phi}$ we say that M is a *ϕ-reduct* of M' and that M' is a *ϕ-expansion* of M.

First-order logic constitutes a long-established example of an institution [10].

Example (Many-sorted first-order logic with equality(FOL)). Signatures. A (many-sorted) first-order signature (S, F, P) consists of a set S of *sorts*, a family F of sets $F_{w\to s}$, for $w \in S^*$ and $s \in S$, of *operation* symbols with arity w and sort s (when the arity is empty, $F_{\lambda\to s}$ denotes the set of *constants* of sort s), and a family P of sets P_w, for $w \in S^*$, of *relation* symbols with arity w indexed by arities. A signature (S, F, P) is *algebraic* when P is empty.

Signature Morphisms. The morphisms of signatures $\phi\colon (S, F, P) \to (S', F', P')$ consist of functions $\phi^{\mathrm{st}}\colon S \to S'$ between the sets of sorts, $\phi^{\mathrm{op}}_{w\to s}\colon F_{w\to s} \to F'_{\phi^{\mathrm{st}}(w)\to\phi^{\mathrm{st}}(s)}$, for $w \in S^*$ and $s \in S$, between the sets of operation symbols, and $\phi^{\mathrm{rel}}_{w}\colon P_w \to P'_{\phi^{\mathrm{st}}(w)}$, for $w \in S^*$, between the sets of relation symbols.

Models. For every signature (S, F, P), a model M interprets every sort symbol s as a set M_s, called the *carrier set of sorts*, every operation symbol $\sigma \in F_{w\to s}$ as a function $M_\sigma\colon M_w \to M_s$, where $M_w = M_{s_1} \times \cdots \times M_{s_n}$ for $w = s_1 \ldots s_n$, with $s_1, \ldots, s_n \in S$, and every relation symbol $\pi \in P_w$ as a subset $M_\pi \subseteq M_w$.

A homomorphism of (S, F, P)-models $h\colon M \to N$ is an indexed family of functions $\{h_s\colon M_s \to N_s \mid s \in S\}$ such that

- h is an (S, F)-algebra homomorphism, that is $h_s(M_\sigma(m)) = N_\sigma(h_w(m))$, for every $\sigma \in F_{w\to s}$ and $m \in M_w$, where $h_w\colon M_w \to N_w$ is the canonical component-wise extension of h to tuples, and
- $h_w(m) \in N_\pi$ if $m \in M_\pi$, i.e., $h_w(M_\pi) \subseteq N_\pi$, for every $\pi \in P_w$.

Model Reducts. For every signature morphism $\phi\colon \Sigma \to \Sigma'$, the *reduct* $M'\!\upharpoonright_{\phi}$ of a Σ'-model M' is defined by $(M'\!\upharpoonright_{\phi})_{\alpha} = M'_{\phi(\alpha)}$ for every sort, function, or relation symbol α from Σ. The reducts of homomorphisms are defined likewise.

Sentences. The sentences are usual first-order sentences built from equational and relational atoms by applying in an iterative manner Boolean connectives and first-order quantifiers. The existential and universal quantification are over sets of first-order variables, which are triples $\langle x, s(S, F, P)\rangle$ sometimes denoted by $x : s$, where x is the name of the variable and $s \in S$ is its sort: for any $(S, F \uplus X, P)$-sentence ρ, $\exists X.\rho$ and $\forall X.\rho$ are (S, F, P)-sentences, where $(S, F \uplus X, P)$ denotes the extension of (S, F, P) with the elements of X as new symbols of constants. Note that different variables in X should have different names.

Sentence Translations. Every signature morphism $\phi\colon (S, F, P) \to (S', F', P')$ induces a sentence translation $\mathrm{Sen}(\phi)\colon \mathrm{Sen}(S, F, P) \to \mathrm{Sen}(S', F', P')$ that is defined inductively on the structure of sentences and renames the symbols of (S, F, P) according to ϕ. For instance, the translation of an existentially quantified sentence is $\mathrm{Sen}(\phi)(\exists X.\rho) = \exists X^{\phi}.\mathrm{Sen}(\phi^X)(\rho)$, where $X^{\phi} = \{x : \phi^{\mathrm{st}}(s) \mid x : s \in X\}$ and $\phi^X : (S, F \uplus X, P) \to (S', F' \uplus X^{\phi}, P')$ extends ϕ canonically.

Satisfaction. The satisfaction relation between models and sentences is the usual Tarskian satisfaction defined inductively on the structure of sentences. For existentially quantified sentences, for example, given a model M of a signature (S, F, P), $M \models \exists X.\rho$ if and only if there exists an expansion M' of M along the signature inclusion $(S, F, P) \hookrightarrow (S, F \uplus X, P)$ such that $M' \models \rho$.

Model Amalgamation. Model amalgamation will prove to be crucial in adding quantifiers over an arbitrary institution. Essentially, it allows us to combine models of different signatures whenever they are compatible with respect to a common sub-signature. Many logical systems of interest for specification theory have model amalgamation, including the examples considered in this paper.

Definition 2. *In any institution, a commuting square of signature morphisms*

$$\begin{array}{ccc} \Sigma & \xrightarrow{\varphi_1} & \Sigma_1 \\ {\scriptstyle \varphi_2}\downarrow & & \downarrow{\scriptstyle \theta_1} \\ \Sigma_2 & \xrightarrow[\theta_2]{} & \Sigma' \end{array}$$

is a weak amalgamation square *if, for each Σ_1-model M_1 and Σ_2-model M_2 such that* $\mathrm{Mod}(\varphi_1)(M_1) = \mathrm{Mod}(\varphi_2)(M_2)$*, there exists a Σ'-model M', called an* amalgamation *of M_1 and M_2, such that* $\mathrm{Mod}(\theta_1)(M') = M_1$ *and* $\mathrm{Mod}(\theta_2)(M') = M_2$*. When M' is unique, the above square is called an* amalgamation square.

We say that an institution has (weak) model amalgamation if and only if each pushout square of signature morphisms is a (weak) amalgamation square.

Therefore, in order to have model amalgamation, the square of signature morphisms must not identify entities of Σ_1 and Σ_2 that do not come from Σ via the signature morphisms φ_1 and φ_2. Moreover, to guarantee the uniqueness of the amalgamation, Σ' must contain only entities that come from Σ_1 or Σ_2.

Presentations. The presentations over an institution represent one of the simplest forms of specifications over that logic being formed merely of a signature and a (usually finite) set of its sentences. We will use presentations in our paper to encode reachability logic into first-order logic.

Definition 3. *The* presentations *of an institution* $\mathbf{I} = (\mathbb{S}\mathrm{ig}, \mathrm{Sen}, \mathrm{Mod}, \models)$ *are pairs* (Σ, E) *consisting of a signature* Σ *and a set* E *of* Σ*-sentences. They form a category* $\mathbb{P}\mathrm{res}$ *whose arrows* $\phi\colon (\Sigma, E) \to (\Sigma', E')$ *are signature morphisms* $\phi\colon \Sigma \to \Sigma'$ *such that* $E' \models \phi(E)$*. By extending the sentence functor, the model functor and the satisfaction relation from the signatures of* $\mathbf{I}$ *to presentations we obtain an institution* $\mathbf{I}^{\mathrm{pres}} = (\mathbb{P}\mathrm{res}, \mathrm{Sen}^{\mathrm{pres}}, \mathrm{Mod}^{\mathrm{pres}}, \models^{\mathrm{pres}})$ *of* $\mathbf{I}$ *presentations.*

Moving Between Institutions. In order to use institutions as formalisations of logical systems in a heterogeneous setting, one needs to define formally a notion of map between institutions. Several concepts have been defined over the years, including semi-morphisms, morphisms, and comorphisms, some of which can be found in [23]. In our work, we focus only on comorphisms [15,24], which reflect the intuition of embedding simpler institutions into more complex ones.

Definition 4. *Given two institutions* $\mathbf{I}$ *and* $\mathbf{I}'$*, a* comorphism $(\Phi, \alpha, \beta)\colon \mathbf{I} \to \mathbf{I}'$ *consists of*

- *a signature functor* $\Phi\colon \mathbb{S}\mathrm{ig} \to \mathbb{S}\mathrm{ig}'$,
- *a natural transformation* $\alpha\colon \mathrm{Sen} \Rightarrow \Phi\,;\mathrm{Sen}'$, *and*
- *a natural transformation* $\beta\colon \Phi^{\mathrm{op}}\,; \mathrm{Mod}' \Rightarrow \mathrm{Mod}$

such that the following satisfaction condition *holds for any* $\mathbf{I}$*-signature* Σ*,* $\Phi(\Sigma)$*-model* M'*, and* Σ*-sentence* ρ*:* $M' \models^{\mathbf{I}'}_{\Phi(\Sigma)} \alpha_\Sigma(\rho)$ *iff* $\beta_\Sigma(M') \models^{\mathbf{I}}_{\Sigma} \rho$.

2.2 $\mathbb{K}$ Semantic Framework

The $\mathbb{K}$ framework [20] is an executable semantic framework based on rewriting and used for defining programming languages, computational calculi, type systems and formal analysis tools. It was developed as an alternative to the existing operational-semantics (SOS) frameworks and has been employed to define actual programming languages such as C [9], Python [12], and Java [3].

In defining semantics for programming languages, $\mathbb{K}$ handles cell-like structures named *configurations* and relies on computational structures – *computations* – to model transitions between these configurations by applying local rewriting *rules*. To illustrate how language specifications can be written in the $\mathbb{K}$ semantic framework, we consider the following running example of the (partial) definition of IMP [1], an elementary imperative programming language.

Listing 1.1. The IMP programming language

```
module IMP-SYNTAX
  syntax AExp ::= Int | Id
                | AExp "/" AExp                 [left, strict]
                > AExp "+" AExp                 [left, strict]
                | "(" AExp ")"                  [bracket]
  syntax BExp ::= Bool
                | AExp "<=" AExp                [seqstrict]
                | "!" BExp                      [strict]
                > BExp "&&" BExp                [left, strict(1)]
                | "(" BExp ")"                  [bracket]
  syntax Block ::= "{" "}" | "{" Stmt "}"
  syntax Stmt ::= Block
                | Id "=" AExp ";"               [strict(2)]
                | "if" "(" BExp ")"
                  Block "else" Block            [strict(1)]
                | "while" "(" BExp ")" Block
                > Stmt Stmt                     [left]
  syntax Pgm ::= "int" Ids ";" Stmt
  syntax Ids ::= List{Id,","}
endmodule

module IMP
  imports IMP-SYNTAX
  syntax KResult ::= Int | Bool
  configuration <t><k> $PGM:Pgm </k>
                   <state> .Map </state></t>

  rule <k> X:Id => I ...</k> <state>... X |-> I ...</state>
  rule I1:Int / I2:Int => I1 /Int I2 when I2 =/=Int 0
  rule I1:Int + I2:Int => I1 +Int I2
  rule ! T:Bool => notBool T
  rule <k> int (X:Id,Xs:Ids => Xs);_ </k>
       <state> Rho:Map (. => X |->0) </state>
       when notBool(X in keys(Rho))
  ...
endmodule
```

In the following sections we introduce two logics in order to formalize the $\mathbb{K}$ framework. *Matching logic* will be used to define the syntactic constructs – the syntax of the specified programming languages – and the patterns matched in the semantic rules, and to partially capture the semantics of the programming languages by defining the states of the running programs. Subsequently, we build *reachability logic* upon matching logic to capture the semantic rules. Its sentences, defined over the signatures of matching logic, correspond to the rules in the $\mathbb{K}$ modules, while the models represent implementations of programming languages. The language definitions written in $\mathbb{K}$ will thus be seen, leaving aside some parsing instructions without logical interpretation, as formal specifications over reachability logic.

3 Matching Logic

The generality of institutions allows them to accommodate a great variety of logical systems. As a downside however, and as it would be expected for such an abstract notion, certain logics cannot be captured in full detail by institutions; that is, by considering them only as institutions we lose precious information. An example is computation-tree logic, for which we lose the distinction between state and path sentences (which, in fact, do not belong to the sentences of computation-tree logic formalised as an institution).

Matching logic falls into the same category, but this time, we lose the sorts of patterns and the states of models. To palliate this, we extend institutions with notions of classes (for sorts) and stratification of models (for states). The end result – the concept of stratified institution with classes – is obtained as a combination of the institutions with classes described in Definition 5 with the stratified institutions introduced in [2].

3.1 Stratified Institutions with Classes

Definition 5. *An* institution with classes *is a tuple* $(\mathbb{S}\mathrm{ig}, \mathrm{Cls}, \mathrm{Sen}, \kappa, \mathrm{Mod}, \models)$, *where*

- $(\mathbb{S}\mathrm{ig}, \mathrm{Sen}, \mathrm{Mod}, \models)$ *is an institution,*
- $\mathrm{Cls}\colon \mathbb{S}\mathrm{ig} \to \mathbb{S}\mathrm{et}$ *is a functor giving for each signature a set whose elements are called* classes *of that signature, and*
- $\kappa\colon \mathrm{Sen} \Rightarrow \mathrm{Cls}$ *is a natural transformation giving a class for each sentence.*

We will use the notation $\mathrm{Sen}(\Sigma)_c$ *for* $\kappa^{-1}(c)$, $c \in \mathrm{Cls}(\Sigma)$ *to denote the set of* Σ*-sentences of class* c.

Example. An immediate example of an institution with classes is the atomic fragment of equational first-order logic. In this case, Cls is the forgetful functor that maps every signature (S, F) to its underlying set of sorts S, and $\kappa_{(S,F)}$ is the function that assigns to each atom $t = t'$ the common sort of t and t'.

Definition 6. *A* stratified institution with classes *is a tuple* $\underline{\mathrm{I}} = (\mathbb{S}\mathrm{ig}, \mathrm{Cls}, \mathrm{Sen}, \kappa, \mathrm{Mod}, [\![\_]\!], \models)$ *consisting of:*

- *a category* $\mathbb{S}\mathrm{ig}$ *of* signatures *and* signature morphisms,
- *a* class functor $\mathrm{Cls}\colon \mathbb{S}\mathrm{ig} \to \mathbb{S}\mathrm{et}$, *giving for every signature a set of* classes,
- *a* sentence functor $\mathrm{Sen}\colon \mathbb{S}\mathrm{ig} \to \mathbb{S}\mathrm{et}$, *defining for every signature a set of* sentences,
- *a natural transformation* $\kappa\colon \mathrm{Sen} \Rightarrow \mathrm{Cls}$, *associating a class to each sentence,*
- *a* model functor $\mathrm{Mod}\colon \mathbb{S}\mathrm{ig}^{op} \to \mathbb{C}\mathrm{at}$, *defining a category of* models *for every signature,*
- *a* stratification $[\![\_]\!]$ *giving*
  - *for every signature* Σ, *a family of functors* $[\![\_]\!]_{\Sigma,c}\colon \mathrm{Mod}(\Sigma) \to \mathbb{S}\mathrm{et}$, *indexed by classes* $c \in \mathrm{Cls}(\Sigma)$, *and*

- *for every signature morphism* $\phi\colon \Sigma \to \Sigma'$, *a functorial family of natural transformations* $[\![\text{-}]\!]_{\phi,c}\colon [\![\text{-}]\!]_{\Sigma',\mathrm{Cls}(\phi)(c)} \Rightarrow \mathrm{Mod}(\phi)\,;\,[\![\text{-}]\!]_{\Sigma,c}$, *indexed by classes* $c \in \mathrm{Cls}(\Sigma)$, *such that* $[\![M']\!]_{\phi,c}$ *is surjective for every* $M' \in |\mathrm{Mod}(\Sigma')|$, *and*
- *a satisfaction relation between models and sentences, parameterised by model states and classes:* $M \models^{m}_{\Sigma,c} \rho$, *where* Σ *is a signature,* $c \in \mathrm{Cls}(\Sigma)$, $M \in |\mathrm{Mod}(\Sigma)|$, $m \in [\![M]\!]_{\Sigma,c}$, *and* $\rho \in \mathrm{Sen}(\Sigma)_c$

such that the following properties are equivalent:

i. $\mathrm{Mod}(\phi)(M') \models^{[\![M']\!]_{\phi,c}(m')}_{\Sigma,c} \rho$
ii. $M' \models^{m'}_{\Sigma',\mathrm{Cls}(\phi)(c)} \mathrm{Sen}(\phi)(\rho)$,

for every signature morphism $\phi\colon \Sigma \to \Sigma'$, *every class* $c \in \mathrm{Cls}(\Sigma)$, *every model* $M' \in |\mathrm{Mod}(\Sigma')|$, *every state* $m' \in [\![M']\!]_{\Sigma',\mathrm{Cls}(\phi)(c)}$, *and every* $\rho \in \mathrm{Sen}(\Sigma)_c$.

The functoriality of $[\![\text{-}]\!]_{\phi,c}\colon [\![\text{-}]\!]_{\Sigma',\mathrm{Cls}(\phi)(c)} \Rightarrow \mathrm{Mod}(\phi)\,;\,[\![\text{-}]\!]_{\Sigma,c}$ *means that for every signature morphisms* $\phi\colon \Sigma \to \Sigma'$, $\phi'\colon \Sigma' \to \Sigma''$, *every* Σ''*-model* M'', *and every class* $c \in \mathrm{Cls}(\Sigma)$, $[\![M'']\!]_{\phi;\phi',c} = [\![M'']\!]_{\phi',\phi(c)}\,;\,[\![M''\!\upharpoonright_{\phi'}]\!]_{\phi,c}$.

$$[\![M'']\!]_{\Sigma'',\phi'(\phi(c))} \xrightarrow{[\![M'']\!]_{\phi',\phi(c)}} [\![M''\!\upharpoonright_{\phi'}]\!]_{\Sigma',\phi(c)} \xrightarrow{[\![M''\upharpoonright_{\phi'}]\!]_{\phi,c}} [\![(M''\!\upharpoonright_{\phi'})\!\upharpoonright_{\phi}]\!]_{\Sigma,c}$$

$$[\![M'']\!]_{\Sigma'',\phi'(\phi(c))} \xrightarrow{[\![M'']\!]_{\phi;\phi',c}} [\![(M''\!\upharpoonright_{\phi'})\!\upharpoonright_{\phi}]\!]_{\Sigma,c}$$

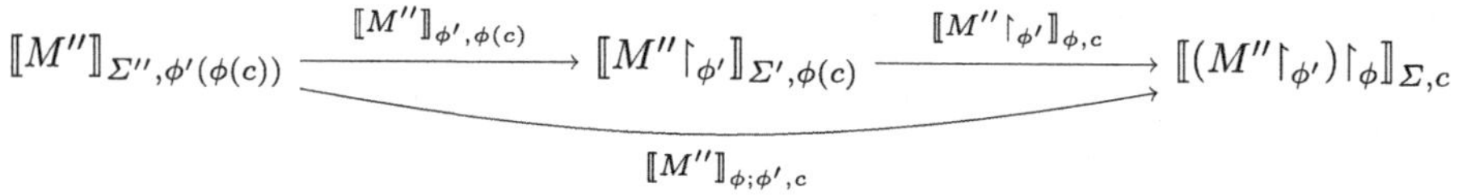

Proposition 1. *Every stratified institution with classes* $\underline{\mathrm{I}} = (\mathbb{S}\mathrm{ig}, \mathrm{Cls}, \mathrm{Sen}, \kappa, \mathrm{Mod}, [\![\text{-}]\!], \models)$ *determines an institution denoted* $\flat\mathbf{I}$ *whose category of signatures is* $\mathbb{S}\mathrm{ig}$, *sentence functor is* Sen, *model functor is* Mod, *and satisfaction relation* $\models_{\Sigma} \subseteq |\mathrm{Mod}(\Sigma)| \times \mathrm{Sen}(\Sigma)$ *is defined, for every signature* $\Sigma \in |\mathbb{S}\mathrm{ig}|$, *as follows:*

$$M \models_{\Sigma} \rho \quad \textit{iff} \quad M \models^{m}_{\Sigma,c} \rho \textit{ for every } m \in [\![M]\!]_{\Sigma,c}, \textit{ where } c = \kappa_{\Sigma}(\rho).$$

Computation-Tree Logic (CTL). We formalise computation-tree logic as a first example of a stratified institution with classes. Similarly to other temporal logics, the usual presentation of CTL is based on propositional logic. CTL inherits the signatures of propositional logic, and thus, the category of its *signatures* is $\mathbb{S}\mathrm{et}$.

CTL formulae can express properties of a state or a path (i.e. an infinite sequence of states) of a transition system (defined below), being *classified* into state and path formulae: $\mathrm{Cls}(\Sigma) = \{\mathit{state}, \mathit{path}\}$, for every $\Sigma \in |\mathbb{S}\mathrm{ig}| = |\mathbb{S}\mathrm{et}|$.

We define the functor Sen, and the natural transformation κ simultaneously, describing the *sentences* of a signature and their classes:

- the atomic propositions $a \in \Sigma$ are sentences of class *state*,
- init is a proposition of class *state*,
- $\varphi_1 \wedge \varphi_2$ is a sentence of class *state*, for every φ_1, φ_2 sentences of class *state*,
- $\neg\varphi$ is a sentence of class *state*, for every sentence φ of class *state*,
- $\exists\pi, \forall\pi$ are sentences of class *state*, for every sentence π of class *path*,
- $\bigcirc\varphi$ is a sentence of class *path*, for every sentence φ of class *state*,
- $\varphi_1 \mathsf{U} \varphi_2$ is a sentence of class *path*, for every φ_1, φ_2 sentences of class *state*.

The *models* of a $\underline{\text{CTL}}$ signature Σ are transition systems $\text{TS} = (S, \rightarrow, I, L)$, where S is a set of states, $\rightarrow \subseteq S \times S$ is a transition relation, $I \subseteq S$ is a set of initial states, and $L\colon S \rightarrow 2^{\Sigma}$ is a labelling function. We define a transition system morphism $h\colon (S, \rightarrow, I, L) \rightarrow (S', \rightarrow', I', L')$ as a function $h\colon S \rightarrow S'$ such that $h(I) \subseteq I'$, $h(\rightarrow) \subseteq \rightarrow'$, and $L(s) = L'(h(s))$, for every $s \in S$.

The *stratification* of models is defined as follows:

- $[\![\text{TS}]\!]_{\Sigma,state}$ is the set S of states of TS,
- $[\![\text{TS}]\!]_{\Sigma,path}$ is the set of paths of TS, that is sequences $s_0, s_1, \ldots \in S^{\omega}$, such that $s_i \rightarrow s_{i+1}$ for $i \in \omega$.

For every signature morphism $\phi\colon \Sigma \rightarrow \Sigma'$, the components of the natural transformations $[\![\_]\!]_{\phi,c}$ are identity functions:

$$[\![\text{TS}']\!]_{\phi,state}(s') = s', \quad [\![\text{TS}']\!]_{\phi,path}(p') = p',$$

for every $s' \in [\![\text{TS}']\!]_{\Sigma',state}$, $p' \in [\![\text{TS}']\!]_{\Sigma',path}$, and $\text{TS}' \in |\text{Mod}(\Sigma')|$.

The *satisfaction relation* between models and sentences is given by:

- $\text{TS} \models^{s}_{\Sigma,state} \rho$ iff
  - $\rho \in L(s)$, for $\rho \in \Sigma$
  - $s \in I$, for $\rho = \text{init}$
  - $\text{TS} \not\models^{s}_{\Sigma,state} \varphi$, for $\rho = \neg\varphi$
  - $\text{TS} \models^{s}_{\Sigma,state} \varphi_1$ and $\text{TS} \models^{s}_{\Sigma,state} \varphi_2$, for $\rho = \varphi_1 \wedge \varphi_2$
  - there is $p = s_0, s_1, \ldots \in [\![\text{TS}]\!]_{\Sigma,path}$ with $s_0 = s$ such that $\text{TS} \models^{p}_{\Sigma,path} \pi$, for $\rho = \exists\pi$,
- $\text{TS} \models^{p}_{\Sigma,path} \rho$ iff
  - $\text{TS} \models^{s_1}_{\Sigma,state} \varphi$, for $\rho = \bigcirc\varphi$
  - there exists an index j such that $\text{TS} \models^{s_j}_{\Sigma,state} \varphi_2$, and for all $i < j$, $\text{TS} \models^{s_i}_{\Sigma,state} \varphi_1$, for $\rho = \varphi_1 \mathsf{U} \varphi_2$,

for every signature Σ, every state $s \in S$, every path $p = s_0, s_1, \ldots$, every sentence ρ and every model $\text{TS} \in |\text{Mod}(\Sigma)|$.

3.2 Matching Logic

The original notion of matching logic developed in [18] can be described as a stratified institution with classes $\underline{\text{ML}}$ as follows.

Signatures. The signatures of $\underline{\text{ML}}$ are algebraic signatures.

Example. Let us consider the specification of the IMP programming language exemplified in Listing 1.1. The signature of the IMP-SYNTAX module is obtained by adding to the built-in syntactic categories and their corresponding semantic operations new sorts and operation symbols introduced by the **syntax** keyword. For example, in the fragment of the syntax module below, the AExp sort is introduced as a supersort of the *Int* and *Id* sorts.[1] Addition and division

[1] For simplicity, the formalism we used in this paper does not take into account the subsorting relation. We could further include subsorts following ideas developed for order-sorted equational logic [11].

are defined as binary operations with arguments and results of sort AExp, while bracketing is defined as a unary operation of the same sort. We note that only the **bracket** attribute has an effect on the signature of the specification as it determines the removal of its corresponding symbol of operation from the signature. The **left** and **strict** attributes are only used in parsing programs and in refining the evaluation strategy (by sequencing computational tasks), and thus, they do not play a role in defining the signature.

Listing 1.2. The IMP programming language – AExp syntax

```
syntax AExp ::= Int | Id
              | AExp "/" AExp          [left, strict]
              > AExp "+" AExp          [left, strict]
              | "(" AExp ")"           [bracket]
```

The signature of the fragment above is $\text{AExp}^{\mathbb{S}\text{ig}} = (S \cup S_{\text{BUILT-IN}}, F \cup F_{\text{BUILT-IN}})$, where $S_{\text{BUILT-IN}}$ and $F_{\text{BUILT-IN}}$ are the built-in sorts and operations, $S = \{\text{AExp}\}$ and $F_{AExp\,AExp \to AExp} = \{\_ + \_, \ \_/\_\}$. Similarly, the signature of the IMP module is obtained from the signature of the imported module IMP-SYNTAX, extending its signature through the addition of the sorts T, K, State and KResult and the operations $\langle k \rangle\_\langle /k \rangle \in F_{\text{Pgm} \to \text{K}}$,[2] $\langle \text{state} \rangle\_\langle /\text{state} \rangle \in F_{\text{Map} \to \text{State}}$ and $\langle \text{t} \rangle\_\langle /\text{t} \rangle \in F_{\text{K State} \to \text{T}}$ introduced by the keyword **configuration**.

Classes of a Signature. Every algebraic signature (S, F) determines (through the functor Cls) the set of classes S, that is the set of its sorts. Similarly, every morphism $\phi\colon (S, F) \to (S', F')$ determines a translation of classes $\phi^{\text{st}}\colon S \to S'$.

Sentences. The sentences (or *patterns*) in ML of given sorts are defined as follows: for every signature Σ, $\text{Sen}(\Sigma)$ is the least set that contains *basic patterns* (terms over Σ, see [18]) and that is closed under the Boolean connectives $\neg, \wedge$, and the existential quantifier $\exists$.

- For each $s \in S$, the basic patterns of sort s are first-order Σ-terms of sort s.
- For every pattern π of sort s, $\neg\pi$ is a pattern of sort s.
- For every two patterns π_1, π_2 of sort s, $\pi_1 \wedge \pi_2$ is a pattern of sort s.
- For every variable x of sort s (defined formally as a tuple $\langle x, s, \Sigma \rangle$, where x is the name of the variable, and s is its sort), and every pattern $\pi \in \text{Sen}(S, F \uplus \{x : s\})$, $\exists x : s.\pi$ is a pattern in $\text{Sen}(\Sigma)$.

The *sentence translation* along a signature morphism $\phi\colon \Sigma \to \Sigma'$ is defined similarly to the translation of first-order sentences. For instance, for basic patterns π of sort s, $\text{Sen}(\phi)(\pi) = \phi^{\text{tm}}_s(\pi)$, where ϕ^{tm} is the extension of ϕ to terms that maps $\sigma(t_1, \ldots, t_n) : s$ to $\phi^{\text{op}}(\sigma)(\phi^{\text{tm}}(t_1), \ldots, \phi^{\text{tm}}(t_n)) : \phi^{\text{st}}(s)$, for every $\sigma \in F_{s_1 \ldots s_n \to s}$, and term t_i of sort s_i.

Example. We can give as examples of sentences of an ML-signature, the patterns matched in the $\mathbb{K}$ rules corresponding to the IMP programming language specification presented in Listing 1.1: I1:*Int* + I2:*Int*, I1:*Int*/I2:*Int* $\wedge$ I2=/=0.

[2] Pgm is the sort defined in Listing 1.1 for capturing the syntax of an IMP program.

Classes of Sentences. The class of a pattern is given by its sort through the natural transformation $\kappa\colon \mathrm{Sen} \Rightarrow \mathrm{Cls}$ that is defined inductively on the structure of sentences:

- $\kappa_{(S,F)}(\pi) = s$, for every basic pattern $\pi \in (T_\Sigma)_s$,
- $\kappa_{(S,F)}(\neg\pi) = \kappa_{(S,F)}(\pi)$, for every pattern π,
- $\kappa_{(S,F)}(\pi_1 \wedge \pi_2) = \kappa_{(S,F)}(\pi_1) = \kappa_{(S,F)}(\pi_2)$, for every two patterns π_1, π_2,
- $\kappa_{(S,F)}(\exists x : s.\pi) = \kappa_{(S,F \uplus \{x\,:\,s\})}(\pi)$, for every pattern π.

Models. The models of ML are *multialgebras* [13]. These are generalisations of algebras having non-deterministic operations that return sets of possible values; that is, multialgebras interpret operation symbols from the carrier set of their arity to the powerset of the carrier set of their sort. For a signature $\Sigma = (S, F)$, a *multialgebra homomorphism* $h\colon M \to N$ is a family of functions indexed by the signature's sorts $\{h_s\colon M_s \to N_s \mid s \in S\}$, such that $h_s(M_\sigma(m_1, \dots, m_n)) \subseteq N_\sigma(h_{s_1}(m_1), \dots, h_{s_n}(m_n))$, for every $\sigma \in F_{s_1 \dots s_n \to s}$ and every $m_i \in M_{s_i}$.

Stratification. The stratification of models is given, for every signature Σ and class s of Σ, by $[\![M]\!]_{\Sigma,s} = M_s$, and for every signature morphism $\phi\colon \Sigma \to \Sigma'$, class s of Σ and model M' of Σ', by $[\![M']\!]_{\phi,s}(m') = m'$, where $m' \in M'_{\phi^{\mathrm{st}}(s)}$.

Satisfaction Relation. The satisfaction relation is based on the interpretation of patterns in models. For any multialgebra M, we define M_π, the interpretation of a pattern π in M, inductively, as follows:

- for every pattern $\pi \in F_{\lambda \to s}$, M_π is the interpretation of the constant π,
- $M_\pi = \bigcup\{M_\sigma(m_1, \dots, m_n) \mid m_i \in M_{t_i}\}$, for every basic pattern $\sigma(t_1, \dots, t_n)$,
- $M_\pi = M_s \setminus M_{\pi_1}$, for every pattern $\pi = \neg\pi_1$, where π_1 is a pattern of sort s,
- $M_\pi = M_{\pi_1} \cap M_{\pi_2}$, for every pattern $\pi = \pi_1 \wedge \pi_2$,
- $M_\pi = \bigcup\{(M, X)_{\pi_1} \mid X \subseteq M_t\}$, for every pattern $\pi = \exists x : t.\pi_1$, where π_1 is a pattern of sort s, and (M, X) is the expansion of M along the inclusion $(S, F) \subseteq (S, F \uplus \{x : t\})$ given by $(M, X)_x = X$.

We now have all the necessary concepts for defining the *satisfaction relation*:

$$M \models^m_{\Sigma,s} \pi \quad \text{iff} \quad m \in M_\pi.$$

Proposition 2. *For every signature morphism $\phi\colon \Sigma \to \Sigma'$, sort $s \in \mathrm{Cls}(\Sigma)$, multialgebra $M' \in \mathrm{Mod}(\Sigma')$, state $m' \in [\![M']\!]_{\Sigma',\mathrm{Cls}(\phi)(s)}$, and pattern π of sort s*

$$\mathrm{Mod}(\phi)(M') \models^{[\![M']\!]_{\phi,s}(m')}_{\Sigma,s} \pi \quad \textit{iff} \quad M' \models^{m'}_{\Sigma',\mathrm{Cls}(\phi)(s)} \mathrm{Sen}(\phi)(\pi).$$

In order to formalise the 𝕂 framework we should interpret the variables in a deterministic manner. For example, in the specification of the IMP language, the variables in the patterns matched by the semantic rules have a deterministic interpretation: the variables I1, I2 and T of the patterns I1 : *Int* + I2 : *Int*, ! T : *Bool* are interpreted as sole elements of sort *Int* or *Bool* respectively, as opposed to the interpretation of variables in ML as sets of elements.

Matching Logic with Deterministic Variables ($\underline{\mathrm{ML}}^+$). We refine the above definition of matching logic $\underline{\mathrm{ML}} = (\mathbb{S}\mathrm{ig}, \mathrm{Cls}, \mathrm{Sen}, \kappa, \mathrm{Mod}, [\![\_]\!], \models)$, by interpreting the variables in a deterministic way, as presented in [18].

$\underline{\mathrm{ML}}^+$ is defined as a stratified institution with classes, whose *category of signatures* is denoted by $\mathbb{S}\mathrm{ig}^+$. Its objects are tuples (S, F, D), where (S, F) and (S, D) are algebraic signatures of $\underline{\mathrm{ML}}$, such that $F_{w \to s} \cap D_{w \to s} = \emptyset$ for every $w \in S^*$, and $s \in S$. For signatures $\Sigma = (S, F, D)$ and $\Sigma' = (S', F', D')$, a signature morphism $\phi \colon \Sigma \to \Sigma'$ is a tuple $(\phi^{\mathrm{st}}, \phi^{\mathrm{op}}, \phi^{\mathrm{det}})$, where the pairs $(\phi^{\mathrm{st}}, \phi^{\mathrm{op}})$ and $(\phi^{\mathrm{st}}, \phi^{\mathrm{det}})$ are signature morphisms in $\mathbb{S}\mathrm{ig}$.

We define the functor $\mathrm{U} \colon \mathbb{S}\mathrm{ig}^+ \to \mathbb{S}\mathrm{ig}$ by $\mathrm{U}(S, F, D) = (S, F \cup D)$ for signatures, and by $\mathrm{U}(\phi) = (\phi^{\mathrm{st}}, \phi^{\mathrm{op}} \cup \phi^{\mathrm{det}})$ for signature morphisms. The *classes* and the *sentences of a signature* are given by the functor compositions $\mathrm{Cls}^+ = \mathrm{U};\mathrm{Cls}$, and $\mathrm{Sen}^+ = \mathrm{U};\mathrm{Sen}$[3] respectively. The *classes of sentences* are determined by the composition $\mathrm{U} \cdot \kappa \colon \mathrm{Sen}^+ \to \mathrm{Cls}^+$ of the functor U with the natural transformation κ, that is, $(U \cdot \kappa)_\Sigma = \kappa_{\mathrm{U}(\Sigma)}$, for every signature Σ.

The *models* of $\underline{\mathrm{ML}}^+$ are determined by the functor $\mathrm{Mod}^+ \colon (\mathbb{S}\mathrm{ig}^+)^{\mathrm{op}} \to \mathbb{C}\mathrm{at}$, that assigns to each signature $\Sigma = (S, F, D)$ the full subcategory of $\mathrm{Mod}(\mathrm{U}(\Sigma))$ consisting of the models M in which every operation symbol in D is interpreted in a deterministic way. For every signature morphism $\phi \colon \Sigma \to \Sigma'$ and every model $M' \in |\mathrm{Mod}^+(\Sigma')|$, we define $\mathrm{Mod}^+(\phi)(M')$ as $\mathrm{Mod}(\mathrm{U}(\phi))(M')$. Notice that the functor Mod^+ is well-defined, as $|\mathrm{Mod}(\mathrm{U}(\phi))(M')_\sigma(m_1, \ldots, m_n)| = |M'_{\phi^{\mathrm{det}}(\sigma)}(m_1, \ldots, m_n)| = 1$ for every operation symbol σ of D.

The *stratification of models* is defined just as in the case of $\underline{\mathrm{ML}}$:

- $[\![\_]\!]^+_{\Sigma,c} \colon \mathrm{Mod}^+(\Sigma) \to \mathbb{S}\mathrm{et}$ maps every model M to $[\![M]\!]_{\mathrm{U}(\Sigma),c}$
- $[\![M]\!]^+_{\phi,c} \colon [\![M]\!]^+_{\Sigma',\mathrm{Cls}^+(\phi)(c)} \to \mathrm{Mod}^+(\phi);[\![M]\!]^+_{\Sigma,c}$ maps every state m to the state $[\![M]\!]_{\mathrm{U}(\phi),c}(m)$, for every morphism $\phi \colon \Sigma \to \Sigma'$ and class c of Σ.

Finally, the *satisfaction relation* between models and sentences is defined analogously to the satisfaction relation of $\underline{\mathrm{ML}}$. As a result, it holds for example, that for any basic pattern π, any signature $\Sigma \in \mathbb{S}\mathrm{ig}^+$, any class $c \in \mathrm{Cls}^+(\Sigma)$, and any model $M \in |\mathrm{Mod}^+(\Sigma)|$, $M(\models^{\underline{\mathrm{ML}}^+})^m_{\Sigma,c}\pi$ iff $M(\models^{\underline{\mathrm{ML}}})^m_{\mathrm{U}(\Sigma),c}\pi$. We note, however, that the satisfaction relation of $\underline{\mathrm{ML}}^+$ is not a restriction of the satisfaction relation of $\underline{\mathrm{ML}}$. For example, if π were an existentially quantified pattern $\exists x\!:\!t.\pi_1$, then only the converse implication of the above equivalence would be ensured to hold. This follows because in $\underline{\mathrm{ML}}$ every expansion of M may interpret in a non-deterministic manner the variable $x\!:\!t$; in order words, there is no guarantee that there exists an expansion of M in $\underline{\mathrm{ML}}$ that satisfies π and is also a model of $\underline{\mathrm{ML}}^+$.

[3] Technically, the quantification in $\underline{\mathrm{ML}}^+$ is done only over variables that are interpreted in a deterministic manner. This means that every extension with variables over signature $\mathrm{U}(\Sigma)$ (in $\underline{\mathrm{ML}}$) corresponds to a deterministic extension of Σ in $\underline{\mathrm{ML}}^+$.

3.3 Encoding Matching Logic into First-Order Logic

There exists a comorphism of institutions between $\flat\mathbf{ML}^+$, the institution obtained from $\underline{\mathrm{ML}}^+$ following Proposition 1, and **FOL**, the institution of first-order logic. In short, the deterministic operations of any given $\flat\mathbf{ML}^+$-signature are preserved by the signature-translation component of the comorphism, while each non-deterministic operation is transformed into a new predicate. In this manner, the interpretation of first-order predicates corresponds to the interpretation of non-deterministic operations in multialgebras. Furthermore, the underlying sentence-translation map of the comorphism encodes each matching pattern to a corresponding universally quantified sentence over states having the same class as the pattern. We define $(\Phi, \alpha, \beta) \colon \flat\mathbf{ML}^+ \to \mathbf{FOL}$ as follows:

For Signatures: The underlying signature functor $\Phi \colon \mathbb{S}\mathrm{ig}^{\flat\mathbf{ML}^+} \to \mathbb{S}\mathrm{ig}^{\mathbf{FOL}}$ maps

- every $\flat\mathbf{ML}^+$ signature $\Sigma = (S, F, D)$ to the **FOL** signature $\Sigma' = (S', F', P')$ where $S' = S$, $F'_{w \to s} = D_{w \to s}$, $P'_\lambda = \emptyset$, and $P'_{ws} = F_{w \to s}$ for $ws \neq \lambda$.
- every $\flat\mathbf{ML}^+$-signature morphism $\phi \colon \Sigma_1 \to \Sigma_2$ to the **FOL**-signature morphism $\phi' = (\phi'^{\mathrm{st}}, \phi'^{\mathrm{op}}, \phi'^{\mathrm{rel}})$, where $\phi'^{\mathrm{st}} = \phi^{\mathrm{st}}$, $\phi'^{\mathrm{op}} = \phi^{\mathrm{det}}$, and $\phi'^{\mathrm{rel}}_{ws} = \phi^{\mathrm{op}}_{w \to s}$, for $ws \neq \lambda$.

For Models: The model functor $\beta_\Sigma \colon \mathrm{Mod}^{\mathbf{FOL}}(\Phi(\Sigma)) \to \mathrm{Mod}^{\flat\mathbf{ML}^+}(\Sigma)$ given by a signature $\Sigma = (S, F, D)$ maps

- every first-order structure M' for $\Phi(\Sigma)$ to the multialgebra M whose carrier sets M_s are defined as M'_s for every sort $s \in S$, whose interpretations $M_\sigma \colon M_{s_1} \times \ldots \times M_{s_n} \to 2^{M_s}$ of function symbols $\sigma \in F_{s_1 \ldots s_n \to s}$ are defined as $M_\sigma(m_1, \ldots, m_n) = \{m \in M_s \mid (m_1, \ldots, m_n, m) \in M'_\sigma\}$, and whose interpretations M_σ of function symbols $\sigma \in D_{w \to s}$ are given by the composition of M'_σ with the singleton-forming map $\{\_\} \colon M_s \to 2^{M_s}$, and
- every morphism of first-order structures $h' \colon M' \to N'$ in $\mathrm{Mod}^{\mathbf{FOL}}(\Phi(\Sigma))$ to a multialgebra morphism $h \colon \beta_\Sigma(M') \to \beta_\Sigma(N')$ given by $h_s = h'_s$, for every $s \in S$. We note that the fact that h' commutes with the interpretation of operation symbols suits the deterministic nature of the morphism h for the interpretation of operations in D, while its compatibility with the interpretation of predicate symbols guarantees the satisfaction of the morphism condition for multialgebras.

For Sentences: The sentence translation $\alpha_\Sigma \colon \mathrm{Sen}^{\flat\mathbf{ML}^+}(\Sigma) \to \mathrm{Sen}^{\mathbf{FOL}}(\Phi(\Sigma))$ given by a signature $\Sigma = (S, F, D)$ maps every Σ-pattern π to the sentence $\alpha_\Sigma(\pi) = \forall m : s.\mathrm{FOL}^{m:s}_\Sigma(\pi)$, where $s = \kappa_\Sigma(\pi)$, m is a first-order variable of sort s for the signature $\Phi(\Sigma)$, and $\mathrm{FOL}^{m:s}_\Sigma \colon \kappa^{-1}_\Sigma(s) \to \mathrm{Sen}^{\mathbf{FOL}}(\Phi(\Sigma) \uplus \{m : s\})$ is the sorted translation of sentences defined as follows:

We begin with a notation: for every operation symbol $\sigma \in (F \cup D)_{s_1, \ldots, s_n \to s}$, and every variables $m_i : s_i$ and $m : s$, we denote by $\sigma^=(m_1, \ldots, m_n, m)$ either the relational atom $\sigma(m_1, \ldots, m_n, m)$ if $\sigma \in F$, or the equational atom $\sigma(m_1, \ldots, m_n) = m$ if $\sigma \in D$.

- for every basic pattern $\pi \in (F \cup D)_{\lambda \to s}$, $\mathrm{FOL}^{m:s}_{\Sigma}(\pi) = \pi^{=}(m)$,
- for every basic pattern $\pi = \sigma(t_1, \ldots, t_n)$, with $\sigma \in (F \cup D)_{s_1,\ldots,s_n \to s}$,

$$\mathrm{FOL}^{m:s}_{\Sigma}(\pi) = \exists m_1 : s_1 \ldots \exists m_n : s_n . \mathrm{FOL}^{m_1:s_1}_{\Sigma}(t_1) \wedge \ldots \wedge \mathrm{FOL}^{m_n:s_n}_{\Sigma}(t_n) \wedge \sigma^{=}(m_1, \ldots, m_n, m),$$

- for every pattern $\pi = \neg\pi_1$, $\mathrm{FOL}^{m:s}_{\Sigma}(\neg\pi_1) = \neg\mathrm{FOL}^{m:s}_{\Sigma}(\pi_1)$,
- for every pattern $\pi = \pi_1 \wedge \pi_2$, $\mathrm{FOL}^{m:s}_{\Sigma}(\pi_1 \wedge \pi_2) = \mathrm{FOL}^{m:s}_{\Sigma}(\pi_1) \wedge \mathrm{FOL}^{m:s}_{\Sigma}(\pi_2)$,
- for every pattern $\pi = \exists x : t.\pi_1$, where $\pi_1 \in \mathrm{Sen}^{\flat\mathbf{ML}^+}(\Sigma \uplus \{x : t\})$, we have $\mathrm{FOL}^{m:s}_{\Sigma}(\exists x : t.\pi_1) = \exists x : t.\xi_{\Sigma}(\mathrm{FOL}^{m:s}_{\Sigma \uplus \{x:t\}}(\pi_1))$, where ξ_Σ is a first-order signature morphism from $\Phi(\Sigma \uplus \{x : t\}) \uplus \{m : s\}$ to $\Phi(\Sigma) \uplus \{m : s\} \uplus \{x : t\}$ defined as the extension of $1_{\Phi(\Sigma)}$ that maps the matching-logic variable $x : t$ for the signature Σ to the first-order variable $x : t$ for the signature $\Phi(\Sigma) \uplus \{m : s\}$, and the first-order variable $m : s$ for the signature $\Phi(\Sigma \uplus \{x : t\})$ to the first-order variable $m : s$ but for the signature $\Phi(\Sigma)$.[4]

The naturality of α results from an analogous property for $\mathrm{FOL}^{m:s}_{\Sigma}$.

Proposition 3. *For every two* $\flat\mathbf{ML}^+$ *signatures* Σ_1, Σ_2, *signature morphism* $\phi \colon \Sigma_1 \to \Sigma_2$, *and variable* $m : s$ *for* Σ_1, *the following diagram commutes.*

$$\begin{array}{ccc}
\kappa^{-1}_{\Sigma_1}(s) & \xrightarrow{\mathrm{FOL}^{m:s}_{\Sigma_1}} & \mathrm{Sen}^{\mathbf{FOL}}(\Phi(\Sigma_1) \uplus \{m : s\}) \\
\Big\downarrow {\scriptstyle \mathrm{Sen}^{\flat\mathbf{ML}^+}(\phi)(_)} & & \Big\downarrow {\scriptstyle \mathrm{Sen}^{\mathbf{FOL}}(\Phi(\phi)^m)} \\
\kappa^{-1}_{\Sigma_2}(\phi^{\mathrm{st}}(s)) & \xrightarrow[\mathrm{FOL}^{m:\phi^{\mathrm{st}}(s)}_{\Sigma_2}]{} & \mathrm{Sen}^{\mathbf{FOL}}(\Phi(\Sigma_2) \uplus \{m : \phi^{\mathrm{st}}(s)\})
\end{array}$$

Satisfaction Condition. In order to show that the definitions of the components of the comorphism given above guarantee that the satisfaction condition holds, it suffices to know that Proposition 4 holds.

Proposition 4. *For every* $\flat\mathbf{ML}^+$ *signature* Σ, *every first-order structure* M *for* $\Phi(\Sigma)$, *and every* Σ*-pattern* π *of sort* s, $M_{\mathrm{FOL}^{m:s}_{\Sigma}(\pi)} = \beta_\Sigma(M)_\pi$.

This can be easily shown by induction on the structure of π, starting with the base case of patterns $\pi \in F_{\lambda \to s}$, for which $M_{\mathrm{FOL}^{m:s}_{\Sigma}(\pi)}$ is the set of states $m \in M_s$ such that $(M, m) \models \pi(m)$, that is M_π, and, by definition, $\beta_\Sigma(M)_\pi = M_\pi$.

4 Reachability Logic

In order to capture reachability logic [21] as an institution, we first define an abstract, parameterised institution over an arbitrary stratified institution with classes, which necessarily has to enjoy properties such as the existence of a

[4] We recall from the definitions of the institutions of matching and first-order logic that from a technical point of view, variables are triples, consisting of name, sort, and signature over which they are defined. Consequently, the signature morphism ξ_Σ maps $\langle x, t, \Sigma \rangle$ to $\langle x, t, \Phi(\Sigma) \uplus \langle m, s, \Phi(\Sigma) \rangle \rangle$, and $\langle m, s, \Phi(\Sigma \uplus x) \rangle$ to $\langle m, s, \Phi(\Sigma) \rangle$.

quantification space, model amalgamation, and preservation of pushouts by the class functor. We then obtain the concrete version of reachability logic that underlies the $\mathbb{K}$ framework by instantiating the parameter of the abstract version with $\underline{\mathrm{ML}}^+$, the stratified institution with classes of matching logic, which we show to satisfy the desired properties.

4.1 Abstract Reachability Logic

We formalise reachability logic in two steps: we begin by describing a sub-institution of reachability logic whose sentences are all atomic (reachability atoms), and we subsequently extend it by adding logical connectives and quantifiers through a general universal-quantification construction.

To define atomic abstract reachability logic we first describe it as a pre-institution [22] whose construction is based upon a stratified institution with classes. This amounts to defining the same elements as those comprised by an institution but without imposing the requirement of the satisfaction condition.

Throughout this section we assume an arbitrary, but fixed stratified institution with classes $\underline{\mathrm{M}} = (\mathbb{S}\mathrm{ig}^{\underline{\mathrm{M}}}, \mathrm{Cls}^{\underline{\mathrm{M}}}, \mathrm{Sen}^{\underline{\mathrm{M}}}, \mathrm{Mod}^{\underline{\mathrm{M}}}, [\![\_]\!]^{\underline{\mathrm{M}}}, \models^{\underline{\mathrm{M}}})$. This serves as a parameter for all the constructions below.

Signatures. The category of signatures of atomic abstract reachability logic, denoted by $\mathbb{S}\mathrm{ig}^{\mathbf{ARL}(\underline{\mathrm{M}})}$, is the same as the category of signatures of $\underline{\mathrm{M}}$.

Sentences. For every signature Σ, $\mathrm{Sen}^{\mathbf{ARL}(\underline{\mathrm{M}})}(\Sigma)$ is the set of pairs of sentences of the stratified institution with classes, denoted by $\pi_1 \Rightarrow \pi_2$, where $\pi_1, \pi_2 \in \mathrm{Sen}^{\underline{\mathrm{M}}}(\Sigma)$. The translation of such a sentence $\pi_1 \Rightarrow \pi_2$ along a signature morphism $\phi\colon \Sigma \to \Sigma'$ is defined as the pair of its translated components according to $\mathrm{Sen}^{\underline{\mathrm{M}}}(\phi)$: $\mathrm{Sen}^{\mathbf{ARL}(\underline{\mathrm{M}})}(\phi)(\pi_1 \Rightarrow \pi_2) = \mathrm{Sen}^{\underline{\mathrm{M}}}(\phi)(\pi_1) \Rightarrow \mathrm{Sen}^{\underline{\mathrm{M}}}(\phi)(\pi_2)$.

Example. If we instantiate the parameter $\underline{\mathrm{M}}$ with the stratified institution with classes $\underline{\mathrm{ML}}^+$, the sentences of $\mathbf{ARL}(\underline{\mathrm{ML}}^+)$ will only capture atomic $\mathbb{K}$ semantic rules, i.e. without quantification and side conditions. This means we could only express atomic rules in the specification of the simple imperative programming language IMP, like **rule** ! true => $\mathrm{not}_{\mathrm{Bool}}$ true.

Models. The reachability models of a signature Σ, given by the $\mathrm{Mod}^{\mathbf{ARL}(\underline{\mathrm{M}})}$ functor, are pairs $(M, \rightsquigarrow)$ of Σ-models M of the underlying stratified institution with classes, and families of preorders $\rightsquigarrow_c \subseteq [\![M]\!]_{\Sigma,c} \times [\![M]\!]_{\Sigma,c}$ indexed by the classes of the signature. The model homomorphisms $h\colon (M_1, \rightsquigarrow_1) \to (M_2, \rightsquigarrow_2)$ are defined as the morphisms between the $\underline{\mathrm{M}}$-models M_1 and M_2 that preserve the preorders: for every $c \in \mathrm{Cls}(\Sigma)$, the function $[\![h]\!]_{\Sigma,c}$ from $([\![M_1]\!]_{\Sigma,c}, \rightsquigarrow_1)$ to $([\![M_2]\!]_{\Sigma,c}, \rightsquigarrow_2)$ is monotone. This allows $\mathrm{Mod}^{\mathbf{ARL}(\underline{\mathrm{M}})}(\Sigma)$ to inherit the identities and the composition of model homomorphisms of $\mathrm{Mod}^{\underline{\mathrm{M}}}(\Sigma)$.

The model reduct $\mathrm{Mod}^{\mathbf{ARL}(\underline{\mathrm{M}})}(\phi)\colon \mathrm{Mod}^{\mathbf{ARL}(\underline{\mathrm{M}})}(\Sigma') \to \mathrm{Mod}^{\mathbf{ARL}(\underline{\mathrm{M}})}(\Sigma)$ given by a signature morphism $\phi\colon \Sigma \to \Sigma'$ is defined as

- $\mathrm{Mod}^{\mathbf{ARL}(\underline{\mathrm{M}})}(\phi)(M', \rightsquigarrow') = (\mathrm{Mod}^{\underline{\mathrm{M}}}(\phi)(M'), \rightsquigarrow)$ for every Σ'-model $(M', \rightsquigarrow')$, where $\rightsquigarrow_c \subseteq [\![M' \restriction_\phi]\!]_{\Sigma,c} \times [\![M' \restriction_\phi]\!]_{\Sigma,c}$ is the reflexive and transitive closure of $[\![M']\!]_{\phi,c}(\rightsquigarrow'_{\phi(c)})$, which will be further denoted by $\to_c$,

- $\mathrm{Mod}^{\mathbf{ARL}(\underline{\mathrm{M}})}(\phi)(h')$ is simply $\mathrm{Mod}^{\underline{\mathrm{M}}}(\phi)(h')$ for every two Σ'-models M'_1, M'_2, and every model homomorphism $h' \colon (M'_1, \rightsquigarrow'_1) \to (M'_2, \rightsquigarrow'_2)$.

Satisfaction Relation. The satisfaction relation between any model $(M, \rightsquigarrow)$ and any sentence $\pi_1 \Rightarrow \pi_2$ is defined as follows: $(M, \rightsquigarrow) \models_{\Sigma}^{\mathbf{ARL}(\underline{\mathrm{M}})} \pi_1 \Rightarrow \pi_2$ if and only if for every $m \in [\![M]\!]_{\Sigma,c}$ such that $M(\models^{\underline{\mathrm{M}}})^m_{\Sigma,c}\pi_1$, there exists $n \in [\![M]\!]_{\Sigma,c}$ such that $M(\models^{\underline{\mathrm{M}}})^n_{\Sigma,c}\pi_2$, and $m \rightsquigarrow_c n$.

Corollary 1. $\mathbf{ARL}(\underline{\mathrm{M}}) = (\mathbb{S}\mathrm{ig}^{\mathbf{ARL}(\underline{\mathrm{M}})}, \mathrm{Sen}^{\mathbf{ARL}(\underline{\mathrm{M}})}, \mathrm{Mod}^{\mathbf{ARL}(\underline{\mathrm{M}})}, \models^{\mathbf{ARL}(\underline{\mathrm{M}})})$ *is a pre-institution.*

The direct implication of the satisfaction condition holds unconditionally.

Proposition 5. *For every signature morphism* $\phi \colon \Sigma \to \Sigma'$, *every class* $c \in \mathrm{Cls}(\Sigma)$, *every model* $(M', \rightsquigarrow') \in |\mathrm{Mod}^{\mathbf{ARL}(\underline{\mathrm{M}})}(\Sigma')|$, *and every sentence* $\pi_1 \Rightarrow \pi_2$,

$$(M', \rightsquigarrow') \models_{\Sigma'}^{\mathbf{ARL}(\underline{\mathrm{M}})} \phi(\pi_1 \Rightarrow \pi_2) \textit{ implies } (M', \rightsquigarrow')\!\upharpoonright_{\phi} \models_{\Sigma}^{\mathbf{ARL}(\underline{\mathrm{M}})} \pi_1 \Rightarrow \pi_2.$$

The converse of Proposition 5 holds if the stratification of the underlying institution of $\mathbf{ARL}(\underline{\mathrm{M}})$ satisfies a property similar to that of lifting relations from [7, Chapter 9].

Proposition 6. *If in* $\mathbf{ARL}(\underline{\mathrm{M}})$, *for every signature morphism* $\phi \colon \Sigma \to \Sigma'$, *every class* $c \in \mathrm{Cls}(\Sigma)$, *every* Σ'*-model* $(M', \rightsquigarrow')$, *and every states* $m' \in [\![M']\!]_{\Sigma',\phi(c)}$ *and* $n \in [\![M'\!\upharpoonright_{\phi}]\!]_{\Sigma,c}$ *such that* $[\![M']\!]_{\phi,c}(m') \rightsquigarrow_c n$, *there exists* $n' \in [\![M']\!]_{\Sigma',\phi(c)}$ *such that* $m' \rightsquigarrow'_{\phi(c)} n'$ *and* $[\![M']\!]_{\phi,c}(n') = n$, *then*

$$(M', \rightsquigarrow')\!\upharpoonright_{\phi} \models_{\Sigma}^{\mathbf{ARL}(\underline{\mathrm{M}})} \pi_1 \Rightarrow \pi_2 \textit{ implies } (M', \rightsquigarrow') \models_{\Sigma'}^{\mathbf{ARL}(\underline{\mathrm{M}})} \phi(\pi_1 \Rightarrow \pi_2),$$

for every sentence $\pi_1 \Rightarrow \pi_2$.

Corollary 2. *If the stratified institution with classes* $\underline{\mathrm{M}}$ *satisfies the hypothesis of Proposition 6, then* $\mathbf{ARL}(\underline{\mathrm{M}})$ *is an institution.*

In most concrete examples of stratified institutions with classes the natural transformations $[\![M']\!]_{\phi,c}$ of the stratification are bijective, or even identities (see for example the definitions of $\underline{\mathrm{ML}}^+$ and $\underline{\mathrm{CTL}}$). Therefore, the hypothesis of Proposition 6 is usually satisfied, entailing that $\mathbf{ARL}(\underline{\mathrm{M}})$ is an institution.

We have hitherto defined only an atomic fragment of the desired institution of abstract reachability logic. To describe the construction of the institution with universally quantified Horn-clause sentences over the atomic sentences of $\mathbf{ARL}(\underline{\mathrm{M}})$, we use the notion of quantification space originating from [8].

Definition 7. *For any category* $\mathbb{S}\mathrm{ig}$ *a class of arrows* $\mathcal{D} \subseteq \mathbb{S}\mathrm{ig}$ *is called a* quantification space *if, for any* $\chi \colon \Sigma \to \Sigma' \in \mathcal{D}$ *and* $\varphi \colon \Sigma \to \Sigma_1$ *there exists a designated pushout*

$$\begin{array}{ccc} \Sigma & \xrightarrow{\varphi} & \Sigma_1 \\ {\scriptstyle \chi}\downarrow & & \downarrow{\scriptstyle \chi(\varphi)} \\ \Sigma' & \xrightarrow[\varphi[\chi]]{} & \Sigma'_1 \end{array}$$

with $\chi(\varphi) \in \mathcal{D}$ *and such that the horizontal composition of these designated pushouts is also a designated pushout, i.e. for the pushouts in the diagram below*

$$\begin{array}{ccccc} \Sigma & \xrightarrow{\varphi} & \Sigma_1 & \xrightarrow{\theta} & \Sigma_2 \\ {\scriptstyle \chi}\downarrow & & \downarrow{\scriptstyle \chi(\varphi)} & & \downarrow{\scriptstyle \chi(\varphi)(\theta)} \\ \Sigma' & \xrightarrow[\varphi[\chi]]{} & \Sigma'_1 & \xrightarrow[\theta[\chi(\varphi)]]{} & \Sigma'_2 \end{array}$$

$\varphi[\chi] \,;\, \theta[\chi(\varphi)] = (\varphi \,;\, \theta)[\chi]$ *and* $\chi(\varphi)(\theta) = \chi(\varphi \,;\, \theta)$, *and such that* $\chi(1_\Sigma) = \chi$ *and* $1_\Sigma[\chi] = 1_{\Sigma'}$. *A quantification space* $\mathcal{D}$ *for* $\mathbb{S}\mathrm{ig}$ *is* adequate *for a functor* Mod: $\mathbb{S}\mathrm{ig}^{\mathrm{op}} \to \mathbb{C}\mathrm{at}$ *when the aforementioned designated pushouts are weak amalgamation squares for* Mod. *A quantification space* $\mathcal{D}$ *for* $\mathbb{S}\mathrm{ig}$ *is called* adequate *for an institution if it is adequate for its model functor.*

Proposition 7. *For any institution* **I** *with an adequate quantification space* $\mathcal{D}$, *the following data defines an institution, called the* institution of universally $\mathcal{D}$-quantified Horn clauses over **I**, *and denoted* **HCL(I)***:*

- $\mathbb{S}\mathrm{ig}^{\mathbf{HCL(I)}} = \mathbb{S}\mathrm{ig}^{\mathbf{I}}$,
- $\mathrm{Mod}^{\mathbf{HCL(I)}} = \mathrm{Mod}^{\mathbf{I}}$,
- $\mathrm{Sen}^{\mathbf{HCL(I)}}(\Sigma)$
  $$= \{\forall\chi.\rho'_1 \wedge \ldots \wedge \rho'_n \to \rho' \mid (\chi\colon \Sigma \to \Sigma') \in \mathcal{D} \text{ and } \rho'_i, \rho' \in \mathrm{Sen}^{\mathbf{I}}(\Sigma')\},$$
  for every signature Σ
- $\mathrm{Sen}^{\mathbf{HCL(I)}}(\varphi)(\forall\chi.\rho'_1 \wedge \ldots \wedge \rho'_n \to \rho')$
  $$= \forall\chi(\varphi).\mathrm{Sen}^{\mathbf{I}}(\varphi[\chi])(\rho'_1) \wedge \ldots \wedge \mathrm{Sen}^{\mathbf{I}}(\varphi[\chi])(\rho'_n) \to \mathrm{Sen}^{\mathbf{I}}(\varphi[\chi])(\rho'),$$
  for every signature morphism $\varphi\colon \Sigma \to \Sigma_1$
- $M \models_\Sigma^{\mathbf{HCL(I)}} \forall\chi.\rho'_1 \wedge \ldots \wedge \rho'_n \to \rho'$
  iff for all χ*-expansions* M' *of* M, $M' \models_{\Sigma'}^{\mathbf{I}} \rho'$ *if* $M' \models_{\Sigma'}^{\mathbf{I}} \rho'_i$ *for* $i = \overline{1,n}$.

To build an institution with universally quantified sentences over $\mathbf{ARL}(\underline{\mathrm{M}})$ as described in Proposition 7, we need to ensure that $\mathbf{ARL}(\underline{\mathrm{M}})$ satisfies its hypothesis. This cannot be guaranteed in general, because $\underline{\mathrm{M}}$ is abstract. Nevertheless, we can obtain an appropriate set of hypotheses for the underlying stratified institution $\underline{\mathrm{M}}$ that allow us to apply Proposition 7:

- the existence of a quantification space for $\mathbf{ARL}(\underline{\mathrm{M}})$ is guaranteed by the existence of a quantification space for $\underline{\mathrm{M}}$, as the categories $\mathbb{S}\mathrm{ig}^{\mathbf{ARL}(\underline{\mathrm{M}})}$ and $\mathbb{S}\mathrm{ig}^{\underline{\mathrm{M}}}$ are equal,

– the fact that $\mathbf{ARL}(\underline{\mathrm{M}})$ has weak model amalgamation follows from the weak model amalgamation property of $\underline{\mathrm{M}}$ (see Definition 8 below) and the preservation of pushouts by the class functor of $\underline{\mathrm{M}}$ (see Proposition 8 below).

Definition 8. *A stratified institution with classes* $\underline{\mathrm{M}}$ *has* (weak) model amalgamation *whenever its corresponding institution* $\flat\mathbf{M}$ *has this property.*

Proposition 8. *For every stratified institution with classes* $\underline{\mathrm{M}}$ *having (weak) model amalgamation such that its class functor* Cls *preserves pushouts,* $\mathbf{ARL}(\underline{\mathrm{M}})$ *has (weak) model amalgamation.*

Corollary 3. *If* $\underline{\mathrm{M}}$ *has an adequate quantification space and a pushout preserving class functor, then* $\mathbf{HCL}(\mathbf{ARL}(\underline{\mathrm{M}}))$ *is an institution.*

4.2 Defining Reachability over Matching Logic

In order to capture reachability logic in its original, concrete form, we must instantiate the parameter of the institution $\mathbf{HCL}(\mathbf{ARL}(\underline{\mathrm{M}}))$ defined above, with the stratified institution $\underline{\mathrm{ML}}^{+}$. To this end, we first point out that by adding variables as deterministic constants to the signatures of $\underline{\mathrm{ML}}^{+}$ we obtain a quantification space. Furthermore, to show that the quantification space is adequate, we use the property of model amalgamation of the comorphism (Φ, α, β) between $\flat\mathbf{ML}^{+}$ and **FOL** defined in Sect. 3.3.

Proposition 9. $\underline{\mathrm{ML}}^{+}$ *has pushouts of signatures. Moreover, its class functor preserves pushouts.*

Example. Let us consider the $\mathbb{K}$ definition of the IMP programming language. By splitting the syntax module into three modules, AExp, BExp and IMP-SYNTAX importing the two expressions modules, we have an immediate and natural example of a pushout of signatures: as both the AExp and BExp modules import the BUILT-IN module containing the built-in sorts and corresponding operations of $\mathbb{K}$, we need to construct the pushout of their signatures in order to obtain the signature of the module IMP-SYNTAX.

$$\begin{array}{ccc} \text{BUILT-IN}^{\mathbb{S}\mathrm{ig}} & \xrightarrow{\subseteq} & \text{AExp}^{\mathbb{S}\mathrm{ig}} \\ {\scriptstyle\subseteq}\downarrow & & \downarrow{\scriptstyle\subseteq} \\ \text{BExp}^{\mathbb{S}\mathrm{ig}} & \xrightarrow[\subseteq]{} & \text{IMP-SYNTAX}^{\mathbb{S}\mathrm{ig}} \end{array}$$

Proposition 10. *In* $\underline{\mathrm{ML}}^{+}$, *the family of extensions with deterministic constants forms a quantification space.*

The following definition originates from [4].

Definition 9. *An institution comorphism* $(\Phi, \alpha, \beta)\colon \mathbf{I} \to \mathbf{I}'$ has weak model amalgamation *if for every* **I**-*signature morphism* $\varphi\colon \Sigma \to \Sigma'$, *every* Σ'-*model* M', *and every* $\Phi(\Sigma)$-*model* N *such that* $\beta_{\Sigma}(N) = M'\restriction_{\varphi}$, *there exists a* $\Phi(\Sigma')$-*model* N' *such that* $\beta_{\Sigma'}(N') = M'$ *and* $N'\restriction_{\Phi(\varphi)} = N$. *We say that* $(\Phi, \alpha, \beta)\colon \mathbf{I} \to \mathbf{I}'$ has model amalgamation *when* N' *is required to be unique.*

Remark 1. *$\underline{\text{ML}}^+$ has model amalgamation. Let us first note that the comorphism (Φ, α, β) between the institution $\flat\mathbf{ML}^+$ and* **FOL** *defined in the previous section has model amalgamation. This property holds trivially since the model reduction functors β_Σ are isomorphisms of categories, for every signature Σ. As the institution of* **FOL** *also has model amalgamation, we can use a general result of institution theory to deduce that $\flat\mathbf{ML}^+$ has model amalgamation.*

Corollary 4. $\mathbf{HCL}(\mathbf{ARL}(\underline{\text{ML}}^+))$ *is an institution.*

4.3 Encoding Reachability Logic into First-Order Logic

For any institution $\mathbf{ARL}(\underline{\text{M}})$ defined over a stratified institution with classes $\underline{M}$, there exists a comorphism of institutions between $\mathbf{ARL}(\underline{\text{M}})$ and $\mathbf{FOL}^{\text{pres}}$, the institution of presentations over first-order logic, whenever there exists a comorphism of institutions (Φ, α, β) between $\flat\mathbf{M}$ and **FOL** such that:

- the classes of a signature in $\mathbb{S}\text{ig}^{\flat\mathbf{M}}$ are given by the sorts of its translation to **FOL**: $\text{Cls} = \Phi\,;\text{St}$, where St is the forgetful functor $\text{St}\colon \mathbb{S}\text{ig}^{\mathbf{FOL}} \to \mathbb{S}\text{et}$,
- for every signature Σ, $\alpha_\Sigma(\pi) = \forall m : s.\text{FOL}_\Sigma^{m\,:\,s}(\pi)$, for every π of class s,[5]
- for every $N \in |\text{Mod}^{\mathbf{FOL}}(\Phi(\Sigma))|$, and every $s \in \text{Cls}(\Sigma)$, $N_s = [\![\beta_\Sigma(N)]\!]_{\Sigma,s}$.

The signature-translation component of the comorphism encodes the reachability relation through the addition of new preorder predicates and corresponding axioms for each class of the signature. The new predicates determine relations on reachable states that define the preorder-family component of a reachability model. The sentence component of the comorphism translates each reachability statement between two patterns to a sentence that expresses the existence of a reachable state for the target pattern for every state of the source pattern. We define the comorphism $(\Phi^R, \alpha^R, \beta^R)\colon \mathbf{ARL}(\underline{\text{M}}) \to \mathbf{FOL}^{\text{pres}}$ as follows:

For Signatures: The signature functor $\Phi^R\colon \mathbb{S}\text{ig}^{\mathbf{ARL}(\underline{\text{M}})} \to \mathbb{S}\text{ig}^{\mathbf{FOL}^{\text{pres}}}$ maps every signature Σ of $\mathbf{ARL}(\underline{\text{M}})$ to $\Phi^R(\Sigma) = (\text{Reach}(\Sigma), E)$, where

- $\text{Reach}(\Sigma)$ denotes the first-order signature obtained by adding to $\Phi(\Sigma) = (S', F', P')$ a predicate *reach* of arity $s\,s$ for every sort $s \in S'$, and
- E is a set of axioms that define the predicates *reach* as preorders: $\{\forall x : s.\mathit{reach}(x, x), \forall x, y, z : s.\mathit{reach}(x, y) \wedge \mathit{reach}(y, z) \to \mathit{reach}(x, z) \mid s \in S'\}$.

For Sentences: For every signature Σ of $\mathbf{ARL}(\underline{\text{M}})$, the sentence translation function $\alpha_\Sigma^R\colon \text{Sen}^{\mathbf{ARL}(\underline{\text{M}})}(\Sigma) \to \text{Sen}^{\mathbf{FOL}}(\text{Reach}(\Sigma))$ maps every Σ-sentence $\pi_1 \Rightarrow \pi_2$ to $\alpha_\Sigma^R(\pi_1 \Rightarrow \pi_2) = \forall m : s.\text{FOL}_\Sigma^{m\,:\,s}(\pi_1) \to \exists n : s.\text{FOL}_\Sigma^{n\,:\,s}(\pi_2) \wedge \mathit{reach}(m, n)$.

For Models: For every signature Σ of $\mathbf{ARL}(\underline{\text{M}})$, $\beta_\Sigma^R\colon \text{Mod}^{\mathbf{FOL}}(\text{Reach}(\Sigma), E) \to \text{Mod}^{\mathbf{ARL}(M)}(\Sigma)$ is the model functor that maps every first-order structure $N \in$

[5] Note that, in this case, $\text{FOL}_\Sigma^{m\,:\,s}(\pi)$ is just a notation, and it should not be confused with the first-order sentence described in the previous section, for which we would need to instantiate $\underline{M}$ with $\underline{\text{ML}}^+$.

$|\mathrm{Mod}^{\mathbf{FOL}}(\mathrm{Reach}(\Sigma), E)|$ to the model $(M, \rightsquigarrow) \in |\mathrm{Mod}^{\mathbf{ARL}(M)}(\Sigma)|$, given by $M = \beta_\Sigma(N) \in |\mathrm{Mod}^{\underline{\mathrm{M}}}(\Sigma)|$ and $\rightsquigarrow_s = \{(m, n) \mid (N, m, n) \models reach(x, y)\}$, for every sort s. Note that $\rightsquigarrow_s$ is well-defined as $\mathrm{Cls} = \Phi \,;\, \mathrm{St}$ and $N_s = [\![M]\!]_{\Sigma,s}$.

To encode the Horn-clause reachability logic of Corollary 4 (defined over $\underline{\mathrm{ML}}^+$) into first-order logic, it suffices to notice that the comorphism considered in Sect. 3.3, $(\Phi, \alpha, \beta)\colon \flat\mathbf{ML}^+ \to \mathbf{FOL}$, satisfies all of the above requirements, and thus can be extended to a comorphism $(\Phi^R, \alpha^R, \beta^R)\colon \mathbf{ARL}(\underline{\mathrm{ML}}^+) \to \mathbf{FOL}^{\mathrm{pres}}$. This can be further extended to an encoding of $\mathbf{HCL}(\mathbf{ARL}(\underline{\mathrm{ML}}^+))$ into $\mathbf{FOL}^{\mathrm{pres}}$ through the use of a general result about Horn-clause institutions.

Proposition 11. *Let* $\mathbf{I}$ *and* $\mathbf{I}'$ *be institutions equipped with quantification spaces. Every comorphism of institutions* $(\Phi, \alpha, \beta)\colon \mathbf{I} \to \mathbf{I}'$ *that has weak model amalgamation, and for which* Φ *preserves the quantification space of* $\mathbf{I}$*, can be extended to a comorphism of institutions between* $\mathbf{HCL}(\mathbf{I})$ *and* $\mathbf{HCL}(\mathbf{I}')$*.*

5 Conclusions and Future Research

In this work, we proposed an institutional formalisation of the logical systems that underlie the $\mathbb{K}$ semantic framework. These logical systems account for the structural properties of program configurations (through matching logic), and changes of these configurations (through reachability logic).

Our work sets the foundation for integrating the $\mathbb{K}$ semantic framework into heterogeneous institution-based toolsets, allowing us to exploit the combined potential of the $\mathbb{K}$ tool and of other software tools such as the MiniSat solver, the SPASS automated prover or the Isabelle interactive proof assistant. Having both matching and reachability logic defined as institutions allows us to integrate them into the logic graphs of institution-based heterogeneous specification languages such as HetCasl [16]. As an immediate result, the $\mathbb{K}$ framework can inherit the powerful module systems developed for specifications built over arbitrary institutions, with dedicated operators for aggregating, renaming, extending, hiding and parameterising modules. In addition, this will enable us to combine reachability logic and the tool support provided by $\mathbb{K}$ with other logical systems and tools. Towards that end, as a preliminary effort to integrate the $\mathbb{K}$ framework into Hets [17], we described comorphisms from matching and reachability logic to the institution of first-order logic.

Another line of research concerns the development of $\mathbb{K}$ from a purely formal-specification perspective, including for example, studies on modularisation and initial semantics. Within this context, verification can be performed based on the proof systems that have already been defined for $\mathbb{K}$.

References

1. The IMP language. http://www.kframework.org/imgs/releases/k/tutorial/1_k/2_imp/lesson_5/imp.pdf
2. Aiguier, M., Diaconescu, R.: Stratified institutions and elementary homomorphisms. Inf. Process. Lett. **103**(1), 5–13 (2007)

3. Bogdănaş, D., Roşu, G.: K-Java: a complete semantics of Java. In: Proceedings of the 42nd Symposium on Principles of Programming Languages, POPL 2015. ACM (2015)
4. Borzyszkowski, T.: Logical systems for structured specifications. Theor. Comput. Sci. **286**(2), 197–245 (2002)
5. Chiriţă, C.E.: An institutional foundation for the K semantic framework. Master's thesis, University of Bucharest (2014)
6. Şerbănuţă, T.F., Arusoaie, A., Lazar, D., Ellison, C., Lucanu, D., Roşu, G.: The K primer (version 3.3). Electron. Notes Theor. Comput. Sci. **304**, 57–80 (2014)
7. Diaconescu, R.: Institution-independent Model Theory. Studies in Universal Logic. Springer, London (2008). http://books.google.ro/books?id=aEpn60-EDXwC
8. Diaconescu, R.: Quasi-boolean encodings and conditionals in algebraic specification. J. Logic Algebraic Program. **79**(2), 174–188 (2010)
9. Ellison, C., Roşu, G.: An executable formal semantics of C with applications. In: Proceedings of the 39th Symposium on Principles of Programming Languages (POPL 2012), pp. 533–544. ACM (2012)
10. Goguen, J.A., Burstall, R.M.: Institutions: abstract model theory for specification and programming. J. ACM **39**(1), 95–146 (1992)
11. Goguen, J.A., Diaconescu, R.: An Oxford survey of order sorted algebra. Math. Struct. Comput. Sci. **4**(3), 363–392 (1994)
12. Guth, D.: A formal semantics of python 3.3. Master's thesis, University of Illinois at Urbana-Champaign, July 2013
13. Lamo, Y.: The Institution of Multialgebras-a general framework for algebraic software development. Ph.D. thesis, University of Bergen (2003)
14. Lane, S.M.: Categories for the Working Mathematician. Springer, New York (1998). http://books.google.ro/books?id=eBvhyc4z8HQC
15. Meseguer, J.: General logics. In: Ebbinghaus, H.D., Fernandez-Prida, J., Garrido, M., Lascar, D., Artalejo, M.R. (eds.) Logic Colloquium 1987 Proceedings of the Colloquium held in Granada, Studies in Logic and the Foundations of Mathematics, vol. 129, pp. 275–329. Elsevier (1989)
16. Mossakowski, T.: HetCasl-heterogeneous specification. Language summary (2004)
17. Mossakowski, T., Maeder, C., Lüttich, K.: The heterogeneous tool set, Hets. In: Grumberg, O., Huth, M. (eds.) TACAS 2007. LNCS, vol. 4424, pp. 519–522. Springer, Heidelberg (2007)
18. Roşu, G.: Matching logic: a logic for structural reasoning. Technical report, University of Illinois, January 2014. http://hdl.handle.net/2142/47004,
19. Roşu, G., Şerbănuţă, T.F.: An overview of the K semantic framework. J. Log. Algebraic Program. **79**(6), 397–434 (2010)
20. Roşu, G., Şerbănuţă, T.F.: K overview and SIMPLE case study. Electron. Notes Theoret. Comput. Sci. **304**, 3–56 (2014)
21. Roşu, G., Ştefănescu, A., Ciobâcă, Ş., Moore, B.M.: One-path reachability logic. In: Proceedings of the 28th Symposium on Logic in Computer Science (LICS 2013), pp. 358–367. IEEE, June 2013
22. Salibra, A., Scollo, G.: Interpolation and compactness in categories of pre-institutions. Math. Struct. Comput. Sci. **6**(3), 261–286 (1996)
23. Sannella, D., Tarlecki, A.: Foundations of Algebraic Specification and Formal Software Development. Monographs in Theoretical Computer Science. Springer, Heidelberg (2012)
24. Tarlecki, A.: Moving between logical systems. In: Haveraaen, M., Dahl, O.-J., Owe, O. (eds.) Abstract Data Types 1995 and COMPASS 1995. LNCS, vol. 1130, pp. 478–502. Springer, Heidelberg (1996)

A Theoretical Foundation for Programming Languages Aggregation

Ştefan Ciobâcă[1(✉)], Dorel Lucanu[1], Vlad Rusu[2], and Grigore Roşu[1,3]

[1] "Alexandru Ioan Cuza" University, Iaşi, Romania
{stefan.ciobaca,dlucanu}@info.uaic.ro
[2] Inria Lille, Villeneuve-d'ascq, France
vlad.rusu@inria.fr
[3] University of Illinois at Urbana-Champaign, Champaign, USA
grosu@illinois.edu

Abstract. Programming languages should be formally specified in order to reason about programs written in them. We show that, given two formally specified programming languages, it is possible to construct the formal semantics of an aggregated language, in which programs consist of pairs of programs from the initial languages. The construction is based on algebraic techniques and it can be used to reduce relational properties (such as equivalence of programs) to reachability properties (in the aggregated language).

1 Introduction

In this paper we are concerned with the problem of language aggregation: given two programming languages (in some formalism), construct a new language in which programs consist of pairs of programs from the original languages. Furthermore, a program (P, Q) in the aggregated language should behave as if the programs P and Q (in the initial languages) would run interleaved or in parallel.

The main motivation behind the construction of the aggregated language is to be able to reduce reasoning about relational properties of programs (such as the equivalence of two programs P and Q) to reasoning about a single program (the aggregated program (P, Q)). We have shown [4] for example that partial equivalence of programs reduces to partial correctness in an aggregated language. In general, aggregation is important because there are fewer results and tools for relational properties (e.g. equivalence of programs) than single program properties (e.g. partial correctness). All of our constructions are effective and therefore aggregation can be implemented as a module in a language framework such as K [15].

The main difficulty in aggregating two languages is making sure that there is a link between the datatypes being shared by the two languages. For example, if both languages have variables of type natural numbers, it is important that the

This paper is supported by the Sectorial Operational Programme Human Resource Development (SOP HRD), financed from the European Social Fund and by the Romanian Government under the contract number POSDRU/159/1.5/S/137750.

M. Codescu et al. (Eds.): WADT 2014, LNCS 9463, pp. 30–47, 2015.
DOI: 10.1007/978-3-319-28114-8_3

naturals be interpreted consistently in the aggregated language in order to be able to express properties such as the equality of a variable in the first language with a variable in the second language. We use known results like pushouts of first-order signatures and amalgamation of first-order models in order to formalize the sharing of information between the two languages. In order to show that the language resulting from the construction is indeed the language we want (i.e. that is has the desirable properties), we have to show several non-trivial results about aggregated configurations (Lemmas 1 and 3) and about the aggregated semantics (Theorems 3, 4 and 5) that were not known before.

In this paper, we show that if the two languages are formalized by their matching logic semantics [11], then the matching logic semantics of the aggregated language can also be constructed. The main advantage of a matching logic semantics is that it allows to faithfully express several operational semantics ([18]) and that a Hoare-like proof system can be obtained for free directly from the semantics [8,14]. Therefore, our method allows one to reason about relational properties (such as equivalence) of programs written in two potentially different programming languages "for free", starting from the matching logic semantics of the languages.

2 Topmost Matching Logic

Matching logic was introduced by Roşu et al. ([11,12]) for specifying programming languages and reasoning about programs. In this section, we recall *topmost matching logic*, a subset of the full matching logic theory described in [13]. For simplicity, we use "matching logic" instead of "topmost matching logic" in this paper.

2.1 Signature

A *matching logic signature* $(\mathit{Cfg}, S, \Sigma, \Pi)$ extends a many-sorted first order signature (S, Σ, Π) (where S is the set of sorts, Σ is the many-sorted set of function symbols and Π is the many-sorted set of predicate symbols) with a sort $\mathit{Cfg} \in S$ of configurations. By Var we denote the (sorted) set of variables. By $T_s(\mathit{Var})$ we denote the set of (well-sorted) terms of sort s built from function symbols in Σ and variables in Var. Matching logic signatures are used to define the abstract syntax of programming languages.

Example 1. The signatures $(\texttt{CfgI}, S_I, \Sigma_I, \Pi_I)$ and $(\texttt{CfgF}, S_F, \Sigma_F, \Pi_F)$ in Fig. 1 model the syntax of an imperative and, respectively, of a functional programming language, with sorts $S_I = \{\texttt{Int}, \texttt{Id}, \texttt{Exp}, \texttt{ExpI}, \texttt{Stmt}, \texttt{Code}, \texttt{CfgI}\}$ in IMP and sorts $S_F = \{\texttt{Id}, \texttt{Int}, \texttt{Exp}, \texttt{ExpF}, \texttt{Val}, \texttt{CfgF}\}$ in FUN, and function symbols

$$\begin{aligned}\Sigma_0 &= \{\_\texttt{+}\_, \_\texttt{-}\_, \_\texttt{*}\_, \_\texttt{/}\_, \_\texttt{<}\_, \_\texttt{==}\_\} \cup \\ &\quad \{\_\texttt{+}\_{\mathit{Int}}, \_\texttt{-}\_{\mathit{Int}}, \_\texttt{*}\_{\mathit{Int}}, \_\texttt{/}\_{\mathit{Int}}, \_\texttt{<}\_{\mathit{Int}}, \_\texttt{<=}\_{\mathit{Int}}, \_\texttt{==}\_{\mathit{Int}}\} \\ \Sigma_I &= \Sigma_0 \cup \{\_\texttt{:=}\_, \texttt{skip}, \_\texttt{;}\_, \texttt{if\_then\_else\_}, \texttt{while\_do\_}, \langle \_, \_ \rangle\} \\ \Sigma_F &= \Sigma_0 \cup \{\_\_, \texttt{letrec\_\_=\_in\_}, \texttt{if\_then\_else\_}, \mu\_.\_, \lambda\_.\_, \langle \_ \rangle\}.\end{aligned}$$

```
   Exp ::= Id | Int | ExpI + ExpI | ExpI - ExpI | ExpI * ExpI | ExpI / ExpI
             | ExpI < ExpI | ExpI <= ExpI | ExpI == ExpI
ExpI ::= Exp                                  ExpF ::= Exp
Stmt ::=  Id := ExpI                                | letrec Id Id = ExpF in ExpF
        | skip | Stmt ; Stmt                        | if ExpF then ExpF else ExpF
        | if ExpI then Stmt else Stmt               | μ Id . ExpF
        | while ExpI do Stmt                        | ExpF ExpF
Code ::=  ExpI | Stmt                         Val  ::=  Int | λ Id . ExpF
CfgI ::= ⟨Code, Map{Id, Int}⟩                 CfgF ::= ⟨ExpF⟩
```

Fig. 1. $(\texttt{CfgI}, S_I, \Sigma_I, \Pi_I)$ and $(\texttt{CfgF}, S_F, \Sigma_F, \Pi_F)$, the signatures of IMP and FUN, detailed in Example 1. Only the function symbols are detailed in the figure; the predicates consist of the arithmetic comparison operators: $\Pi_I = \Pi_F = \{=_{Int}, <_{Int}, \leq_{Int}\}$. The difference between the operators $\_\texttt{+}\_$, $\_\texttt{*}\_$, etc. and their correspondants $\_+\_{}_{Int}$, $\_*\_{}_{Int}$, etc. is that the former are the syntactic language constructs for addition, etc., while the latter are the actual function symbols denoting integer addition, etc.

The functions above are written in Maude-like notation [7], the underscore (_) denoting the position of an argument. Althought not written explicitly above, the signatures also include the one-argument injections needed to inject sorts like `Int` and `Id` into `ExpI`.

2.2 Syntax

Given a matching logic signature (Cfg, S, Σ, Π), the set of *matching logic formulae* is given by the following grammar:

$$\varphi ::= P(t_1, \ldots, t_n), \neg\varphi, \varphi \wedge \varphi, \exists x.\varphi, \pi,$$

where P ranges over Π, $t_1, \ldots, t_n$ are terms of the appropriate sort for the predicate P, $x \in Var$ is a variable and $\pi \in \mathcal{T}_{Cfg}(Var)$ is a term of sort Cfg.

Matching logic formulae include the classical constructs in first-order logic (predicates $P(t_1, \ldots, t_n)$, negation, conjunction and existential quantifier) and also a new construct π, called *basic pattern*, which allows to use terms of sort Cfg as atomic formulae. We also assume that the other first-order connectives (disjunction - $\vee$, implication - $\rightarrow$, universal quantifier - $\forall$) are available and interpreted as usual as syntactic sugar over the existing connectives.

Example 2. The expression

$$\langle \texttt{while}\,(E)\ S,\ \texttt{x} \mapsto a\ \texttt{y} \mapsto b\rangle \wedge a <_{Int} b$$

is a $(\texttt{CfgI}, S_I, \Sigma_I, \Pi_I)$-matching logic formula, where $(\texttt{CfgI}, S_I, \Sigma_I, \Pi_I)$ is that given in Fig. 1.

We distinguish two particular types of matching logic formulae:

Definition 1. *A matching logic formula is* patternless *if it conforms to the following grammar:*

$$\varphi_{pless} ::= P(t_1, \ldots, t_n), \neg\varphi_{pless}, \varphi_{pless} \wedge \varphi_{pless}, \exists x.\varphi_{pless}.$$

Patternless matching logic formulae simply do not contain basic patterns.
Therefore they can be identified with FOL formulae. The second particular type of formulae we consider are *pure* formulae:

Definition 2. *A matching logic formula is* pure *if it conforms to the following grammar:*

$$\varphi_{pure} ::= \pi, \varphi_{pure} \wedge \varphi_{pless}, \exists x.\varphi_{pure}.$$

Pure formulae contain at least one basic pattern and no basic pattern appears under negation.

2.3 Semantics

We denote by $\mathcal{T}$ a first-order model for the many-sorted first-order signature (S, Σ, Π) that assigns sets to sorts, functions to function symbols and predicates to predicate symbols. By $\mathcal{T}_o$ we denote the interpretation of the object o in the model $\mathcal{T}$. Well-sorted valuations are denoted by $\rho : Var \to \mathcal{T}$. Elements of $\mathcal{T}_{Cfg}$ (the set interpreting the sort of configurations) are denoted by the greek letter $\gamma \in \mathcal{T}_{Cfg}$ and are called *configurations.*

Matching logic formulae are interpreted in the presence of a (first-order) model $\mathcal{T}$, a (well-sorted) valuation ρ and an element $\gamma \in \mathcal{T}_{Cfg}$.

Definition 3. *The satisfaction relation* $\models$ *for matching logic is defined as follows:*

1. $\mathcal{T}, \gamma, \rho \models P(t_1, \ldots, t_n)$ *if* $(\rho(t_1), \ldots, \rho(t_n)) \in \mathcal{T}_P$*;*
2. $\mathcal{T}, \gamma, \rho \models \neg\varphi$ *if* $\mathcal{T}, \gamma, \rho \not\models \varphi$*;*
3. $\mathcal{T}, \gamma, \rho \models \varphi_1 \wedge \varphi_2$ *if* $\mathcal{T}, \gamma, \rho \models \varphi_1$ *and* $\mathcal{T}, \gamma, \rho \models \varphi_2$*;*
4. $\mathcal{T}, \gamma, \rho \models \exists x.\varphi$*, where* x *is a variable of sort* s*, if there exists an element* $u \in \mathcal{T}_s$ *such that* $\mathcal{T}, \gamma, \rho[x \mapsto u] \models \varphi$*;*
5. $\mathcal{T}, \gamma, \rho \models \pi$ *for a basic pattern* $\pi \in \mathcal{T}_{Cfg}(Var)$ *if* $\rho(\pi) = \gamma$.

The first four cases are as in first-order logic and the last case (for basic patterns) is new. The semantics of basic patterns is given by *matching* the element γ in the presence of which matching logic formulae are evaluated. This is where the name of *matching logic* comes from. When two basic patterns are connected by a logical and ($\wedge$) in the formula, the element γ has to match both basic patterns, hence $\wedge$ plays the role of intersection. Using basic patterns under implications, logical or and existential/universal quantifiers makes it possible to express several interesting properties, but this is outside the scope of the current article.

Example 3. Let $\mathcal{T}_I$ denote the model for Σ_I that interprets Int as the set of integers, the function and the predicate symbols over Int with the usual functions and predicates, respectively, and the function symbols corresponding to the BNF productions as term constructors. If

$$\gamma = \langle \texttt{while} \quad \texttt{(x<y)} \quad \texttt{x := x+1;}, \texttt{x} \mapsto 2 \ \texttt{y} \mapsto 5 \rangle$$

and $\rho(E) = \texttt{x<y}$, $\rho(S) = \texttt{x := x+1;}$, $\rho(a) = 2$, $\rho(b) = 5$ then

$$\mathcal{T}_I, \gamma, \rho \models \langle \texttt{while} \quad (E) \ S, \texttt{x} \mapsto a \ \texttt{y} \mapsto b \rangle \wedge a <_{Int} b.$$

The Σ_F-model $\mathcal{T}_F$ is defined in a similar way. If

$$\gamma' = \langle \texttt{letrec} \quad \texttt{f x = if} \quad \texttt{(x<1)} \quad \texttt{then} \quad \texttt{1} \quad \texttt{else} \quad \texttt{x*f(x-1)} \quad \texttt{in} \quad \texttt{f(5)} \rangle$$

and $\rho'(F) = \texttt{f}$, $\rho'(X) = \texttt{x}$ and

$$\rho'(E_1) = \texttt{if} \quad \texttt{(x<1)} \quad \texttt{then} \quad \texttt{1} \quad \texttt{else} \quad \texttt{x*f(x-1)}, \rho'(E_2) = \texttt{f(5)},$$

then

$$\mathcal{T}_F, \gamma', \rho' \models \langle \texttt{letrec} \ F \ X = E_1 \ \texttt{in} \ E_2 \rangle \wedge true.$$

3 Reachability Logic

While matching logic allows to reason about individual configurations (of a program), reachability logic builds on matching logic to allow to reason and define the dynamic behaviour of programs.

3.1 Syntax

Reachability logic formulae are constructed in the presence of a matching logic signature (Cfg, S, Σ, Π) as pairs of matching logic formulae:

Definition 4. *A reachability logic formula (or equivalently, a reachability rule) $\varphi \Rightarrow \varphi'$ is a pair of matching logic formulae.*

The intuition behind reachability formulae is that a configuration matching φ advances into a configuration matching φ'.

Example 4. The set of reachability logic formulas $\mathcal{A}_I$ given in Fig. 2 are specifying the semantics of IMP and the set of reachability logic formulas $\mathcal{A}_F$ given in Fig. 3 are specifying the semantics of FUN. The specification

$$\langle C[code], \ \sigma \rangle \Rightarrow \langle C[code'], \ \sigma' \rangle \quad \text{if} \langle code, \ \sigma \rangle \Rightarrow \langle code', \ \sigma' \rangle, \qquad (1)$$

(resp. $\langle C[c] \rangle \Rightarrow \langle C[c'] \rangle \text{if} \langle c \rangle \Rightarrow \langle c' \rangle$) is a rule schemata that defines an infinite set of reachability logic formulas.

$\langle x\texttt{,}\ \sigma\rangle \Rightarrow \langle eval(\sigma, x)\texttt{,}\ \sigma\rangle$
$\langle i_1\ \texttt{op}\ i_2\texttt{,}\ \sigma\rangle \Rightarrow \langle i_1\ \texttt{op}_{Int}\ i_2\texttt{,}\ \sigma\rangle$
$\langle X\ \texttt{:=}\ I\texttt{,}\ \sigma\rangle \Rightarrow \langle \texttt{skip,}\ \sigma[I/X]\rangle$
$\langle \texttt{skip;}S\texttt{,}\ \sigma\rangle \Rightarrow \langle S\texttt{,}\ \sigma\rangle$
$\langle \texttt{if I then}\ S_1\ \texttt{else}\ S_2\texttt{,}\ \sigma\rangle \wedge I \neq 0 \Rightarrow \langle S_1\texttt{,}\ \sigma\rangle$
$\langle \texttt{if 0 then}\ S_1\ \texttt{else}\ S_2\texttt{,}\ \sigma\rangle \Rightarrow \langle S_2\texttt{,}\ \sigma\rangle$
$\langle \texttt{while}\ E\ \texttt{do}\ S\texttt{,}\ \sigma\rangle \Rightarrow \langle \texttt{if}\ E\ \texttt{then}\ S\texttt{;}\ \texttt{while}\ E\ \texttt{do}\ S\ \texttt{else skip,}\ \sigma\rangle$
$\langle C[code]\texttt{,}\ \sigma\rangle \Rightarrow \langle C[code']\texttt{,}\ \sigma'\rangle$ if $\langle code\texttt{,}\ \sigma\rangle \Rightarrow \langle code'\texttt{,}\ \sigma'\rangle$
where $C\ \texttt{::=}\ \_\ |\ C\ \texttt{op}\ E\ |\ i\ \texttt{op}\ C\ |\ \texttt{if}\ C\ \texttt{then}\ S_1\ \texttt{else}\ S_2\ |\ v\ \texttt{:=}\ C\ |\ C\texttt{;}\ S$

Fig. 2. Specifying the semantics of `IMP` as a set $\mathcal{A}_I$ of reachability rules (schemata). *op* ranges over the binary function symbols and op_{Int} is their denotation in $\mathcal{T}_I$.

$\langle I_1 \texttt{op}\, I_2\rangle \Rightarrow \langle I_1 \texttt{op}_{Int} I_2\rangle$
$\langle \texttt{if I then}\ E_1\ \texttt{else}\ E_2\rangle \wedge i \neq 0 \Rightarrow \langle E_1\rangle$
$\langle \texttt{if 0 then}\ E_1\ \texttt{else}\ E_2\rangle \Rightarrow \langle E_2\rangle$
$\langle \texttt{letrec}\ F\ X = E\ \texttt{in}\ E'\rangle \Rightarrow \langle E'[f \mapsto (\mu F.\lambda X.E)]\rangle$
$\langle (\lambda X.E)\ V\rangle \Rightarrow \langle E[V/X]\rangle$
$\langle \mu X.E\rangle \Rightarrow \langle E[X \mapsto (\mu X.E)]\rangle$
$\langle C[c]\rangle \Rightarrow \langle C[c']\rangle$ if $\langle c\rangle \Rightarrow \langle c'\rangle$
where $C\ \texttt{::=}\ \_\ |\ C\ \texttt{op}\ e\ |\ \texttt{if}\ C\ \texttt{then}\ e_1\ \texttt{else}\ e_2\ |\ C\ e\ |\ v\ C$

Fig. 3. Specifying the semantics of `FUN` as a set $\mathcal{A}_F$ of reachability rules schemata. *op* ranges over the binary function symbols and op_{Int} is their denotation in $\mathcal{T}_F$

3.2 Semantics

Reachability logic formulae are interpreted in the presence of a first-order model $\mathcal{T}$ and a (well-sorted) valuation $\rho : Var \to \mathcal{T}$ as in the case of matching logic, and also in the presence of a transition relation $\longrightarrow\ \subseteq \mathcal{T}_{Cfg} \times \mathcal{T}_{Cfg}$ over the set of configurations $\mathcal{T}_{Cfg}$.

Intuitively, a reachability logic formula $\varphi \Rightarrow \varphi'$ holds if any configuration matched by φ reaches (in one step) a configuration matched by the formula φ'. Note that the one-step requirement we work with in this article defines a satisfaction relation that is different from the ones used in the previous presentations [8, 14] of reachability logic. The satisfaction relation defined here is mainly used to specify transition systems.

Definition 5. *Formally, the satisfaction relation* $\models$ *of reachability logic is defined as follows:*

$$\mathcal{T}, \longrightarrow, \rho\ \models\ \varphi\ \Rightarrow \varphi'$$

if for any $\gamma \in \mathcal{T}_{Cfg}$ *such that* $\mathcal{T}, \gamma,\ \rho\ \models\ \varphi$ *there exists a* $\gamma' \in \mathcal{T}_{Cfg}$ *such that* $\gamma \longrightarrow \gamma'$ *and* $\mathcal{T}, \gamma', \rho\ \models\ \varphi'$.

Note that the free variables of φ and the free variables of φ' are "shared" in the reachability formula, in the sense that both are interpreted by the same valuation.

Example 5. If

$$\langle \texttt{if (1) then x=x+y else x := 0}, \texttt{x} \mapsto 2\ \texttt{y} \mapsto 5\rangle \Rightarrow \langle \texttt{x=x+y}, \texttt{x} \mapsto 2\ \texttt{y} \mapsto 5\rangle$$

and $\varphi \Rightarrow \varphi'$ is

$$\langle \texttt{if}\ (I)\ \texttt{then}\ S_1\ \texttt{else}\ S_2, \sigma\rangle \wedge I \neq_{Int} 0 \Rightarrow \langle S_1, \sigma\rangle,$$

then we have $\mathcal{T}_I, \longrightarrow, \rho \models \varphi \Rightarrow \varphi'$, where $\rho(I) = \texttt{1}$, $\rho(S_1) = \texttt{x := x+y}$, $\rho(S_2) = \texttt{x := 0}$, and $\rho(\sigma) = \texttt{x} \mapsto 2\ \texttt{y} \mapsto 5$.

If the valuation ρ is missing, the free variables of reachability rules are interpreted universally:

Definition 6. *Given a first-order model $\mathcal{T}$ and a transition relation $\longrightarrow$, we say that the pair $(\mathcal{T}, \longrightarrow)$ is a model of $\varphi \Rightarrow \varphi'$, written*

$$\mathcal{T}, \longrightarrow \models \varphi \Rightarrow \varphi'$$

if, for all (well-sorted) valuations $\rho : Var \to \mathcal{T}$, we have that $\mathcal{T}, \longrightarrow, \rho \models \varphi \Rightarrow \varphi'$.

The universal interpretation of the free variables is justified in the following section, but note that it is not unusual to do so: for example, the same happens with first-oder clauses, where the variables are (implicitly) universally quantified.

4 Language Semantics

In this section, we show that reachability formulae can be used to formally define the operational semantics of a programming language. We consider that a matching logic signature (Cfg, S, Σ, Π) is fixed.

Definition 7. *A (Cfg, S, Σ, Π)-programming language is a pair $(\mathcal{T}, \longrightarrow)$ of a first-order model $\mathcal{T}$ of (Cfg, S, Σ, Π) and a transition relation $\longrightarrow \subseteq \mathcal{T}_{Cfg} \times \mathcal{T}_{Cfg}$.*

When (Cfg, S, Σ, Π) is understood from the context, we omit it and write programming language instead of (Cfg, S, Σ, Π)-programming language.

The matching logic signature defines the syntax of the language, the model $\mathcal{T}$ mainly defines program configurations and $\longrightarrow$ defines the (one-step) transition relation between configurations. Let A be a set of reachability formulae (the axioms of the language).

Definition 8. *We say that $(\mathcal{T}, \longrightarrow)$ is a* model *of A, and we write $(\mathcal{T}, \longrightarrow) \models A$ if $(\mathcal{T}, \longrightarrow) \models \varphi \Rightarrow \varphi'$ for every reachability formula $\varphi \Rightarrow \varphi' \in A$.*

Example 6. If $(\mathcal{T}_I, \longrightarrow_I)$ gives the semantics of IMP, then we have $(\mathcal{T}_I, \longrightarrow_I) \models \mathcal{A}_I$, i.e., $(\mathcal{T}_I, \longrightarrow_I)$ is a model of $\mathcal{A}_I$. Similarly, if $(\mathcal{T}_F, \longrightarrow_F)$ gives the semantics of FUN, then we must have $(\mathcal{T}_F, \longrightarrow_F) \models \mathcal{A}_F$, i.e., $(\mathcal{T}_F, \longrightarrow_F)$ is a model of $\mathcal{A}_F$.

Therefore, the set A of reachability rules are considered to be the specification (the formal semantics) of any language $(\mathcal{T}, \longrightarrow)$ that is a model of the rules. In [18] it is shown that any operational semantics (small-step SOS, big-step SOS, reduction contexts, etc.) can be faithfully captured by a (possibly infinite) set of reachability rules. Moreover, all such reachability rules from the semantics are *pure*:

Definition 9. *A reachability formula $\varphi \Rightarrow \varphi'$ is* pure *if both φ and φ' are pure.*

From here on, we assume that the formal semantics of any language is given as a (possibly infinite) set of pure reachability formulae.

5 Language Aggregation

We assume two signatures $(Cfg_1, S_1, \Sigma_1, \Pi_1)$ and $(Cfg_2, S_2, \Sigma_2, \Pi_2)$ for two languages $(\mathcal{T}_1, \longrightarrow_1)$ and $(\mathcal{T}_2, \longrightarrow_2)$ specified by the sets A_1 and A_2 of reachability rules.

In this section, we construct the aggregated signature (Cfg, S, Σ, Π) of the aggregated language from the signature of the first language and the signature of the second language. Also, we define the aggregated language $(\mathcal{T}, \longrightarrow)$ itself and show how to constructively give the aggregated axioms A of the new language from the initial languages.

5.1 Signature Aggregation

This subsection is dedicated to showing how to construct the aggregated signature (Cfg, S, Σ, Π) from the individual signature $(Cfg_1, S_1, \Sigma_1, \Pi_1)$ (of the first language) and $(Cfg_2, S_2, \Sigma_2, \Pi_2)$ (of the second language).

The most delicate part is to make sure that the sorts, function and predicate symbols "shared" between the two language are identified in the aggregated configuration, even if their names are not the same in the first and in the second language.

Therefore, we assume that there exists a first-order signature (S_0, Σ_0, Π_0) that the two languages have in common. This means there exist two morphisms $h_1 : (S_0, \Sigma_0, \Pi_0) \to (S_1, \Sigma_1, \Pi_1)$ and $h_2 : (S_0, \Sigma_0, \Pi_0) \to (S_2, \Sigma_2, \Pi_2)$.

Example 7. For the signatures of the two languages, IMP and FUN, we have $S_0 = \{\texttt{Int}, \texttt{Id}, \texttt{Exp}\}$, Σ_0 that was defined on Page 1, and $\Pi_0 = \Pi_1, \Pi_2$. The morphisms h_1 and h_2 are given by the component inclusions.

The following theorem (the pushout theorem) allows us to combine the two signatures into a single signature, while identifying the objects shared between them. This result is not new, see, for example, [9].

Theorem 1 (Pushout of Signatures). *Let (S_1, Σ_1, Π_1), (S_2, Σ_2, Π_2) and (S_0, Σ_0, Π_0) be many-sorted FOL signatures, h_1 a morphism from (S_0, Σ_0, Π_0) to (S_1, Σ_1, Π_1) and h_2 a morphism from (S_0, Σ_0, Π_0) to (S_2, Σ_2, Π_2).*

Then the diagram $(S_1, \Sigma_1, \Pi_1) \xleftarrow{h_1} (S_0, \Sigma_0, \Pi_0) \xrightarrow{h_2} (S_2, \Sigma_2, \Pi_2)$ admits a pushout, i.e., there exists a tuple $(h'_1, (S', \Sigma', \Pi'), h'_2)$ with h'_1 a morphism from (S_1, Σ_1, Π_1) to (S', Σ', Π') and h'_2 a morphism from (S_2, Σ_2, Π_2) to (S', Σ', Π') such that:

1. *(commutativity) $h'_1(h_1(x)) = h'_2(h_2(x))$ for any object x from the signature (S_0, Σ_0, Π_0) and*
2. *(minimality) if there exist (S'', Σ'', Π'') and morphisms h''_1 from (S_1, Σ_1, Π_1) to (S'', Σ'', Π'') and h''_2 from (S_2, Σ_2, Π_2) to (S'', Σ'', Π'') with $h''_1(h_1(x)) = h''_2(h_2(x))$ for all $x \in S_0, \cup \Sigma_0 \cup \Pi_0$, then there exists a morphism h from (S', Σ', Π') to (S'', Σ'', Π'').*

Furthermore, the pushout is unique (up to isomorphisms). The push-out is summarised in Fig. 4.

$$
\begin{array}{ccc}
(S_0, \Sigma_0, \Pi_0) & \xrightarrow{h_2} & (S_2, \Sigma_2, \Pi_2) \\
h_1 \downarrow & & \downarrow h'_2 \\
(S_1, \Sigma_1, \Pi_1) & \xrightarrow{h'_1} & (S', \Sigma', \Pi')
\end{array}
$$

Fig. 4. Push-out diagram assumed throughout the paper.

The first step to obtain the aggregated signature is to apply the push-out theorem in order to obtain the intermediate signature (S', Σ', Π') and the two morphisms $h'_1 : (S_1, \Sigma_1, \Pi_1) \to (S', \Sigma', \Pi')$ and $h'_2 : (S_2, \Sigma_2, \Pi_2) \to (S', \Sigma', \Pi')$.

The first-order signature (S', Σ', Π') contains all of the objects from the initial signatures (properly renamed to account for shared objects), but it does not yet have a sort for aggregated configurations. Let $Cfg' = h_1(Cfg_1)$ and $Cfg'_2 = h_2(Cfg_2)$ be the names of the sorts of configurations in the new signature.

We therefore choose a fresh sort Cfg for aggregated configurations and we let $S = S' \uplus \{Cfg\}$ be the set of sorts. The signature Σ contains, in addition to the symbols in Σ', a pairing symbol and the respective projections. Formally,

$$\Sigma = \Sigma' \uplus \{\langle \_, \_ \rangle : Cfg'_1 \times Cfg'_2 \to Cfg, pr_1 : Cfg \to Cfg'_1, pr_2 : Cfg \to Cfg'_2\}.$$

The pairing symbol $\langle \_, \_ \rangle$ takes as input two configurations of the initial languages and returns a configuration of the aggregated language. The projection operations pr_1 and pr_2 take an aggregated configuration and deconstruct it into the initial configurations. Finally, we let $\Pi = \Pi'$.

The signature (Cfg, S, Σ, Π) is the aggregated signature of the language and its construction is summarised in Fig. 5.

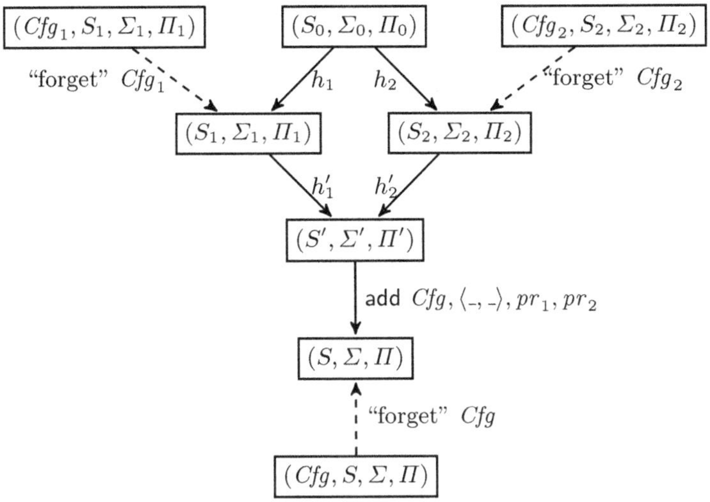

Fig. 5. The aggregation of the two signatures.

If $h : (S, \Sigma, \Pi) \to (S', \Sigma', \Pi')$ is a morphism between the two first order signatures, we extend h to terms as expected: if $t \in T_s(\Sigma)$, then $h(t) \in T_{h(s)}(\Sigma')$. We also extend h to transform matching logic formula in the signature (Cfg, S, Σ, Π) to matching logic formula in the signature $(h(Cfg), S', \Sigma', \Pi')$ as follows:

1. $h(\varphi_1 \wedge \varphi_2) = h(\varphi_1) \wedge h(\varphi_2)$,
2. $h(\exists x.\varphi_1) = \exists x.h(\varphi_1)$,
3. $h(\neg\varphi_1) = \neg h(\varphi_1)$ and
4. $h(P(t_1, \ldots, t_n)) = P(h(t_1), \ldots, h(t_n)))$.

Note that there is no need to have a case for computing $h(\pi)$, since a basic pattern π is nothing but a term. Therefore $h(\pi)$ is already defined. We also extend h to transform reachability formulae over the signature (Cfg, S, Σ, Π) into reachability formulae over the signature $(h(Cfg), S', \Sigma', \Pi')$:

$$h(\varphi \Rightarrow \varphi') = h(\varphi) \Rightarrow h(\varphi').$$

5.2 Model Amalgamation

In this subsection, given two models $\mathcal{T}_1$ and $\mathcal{T}_2$ for the matching logic signatures $(Cfg_1, S_1, \Sigma_1, \Pi_1)$ and $(Cfg_2, S_2, \Sigma_2, \Pi_2)$, we show how to construct a model $\mathcal{T}$ for the aggregated signature (Cfg, S, Σ, Π) above.

In order to construct such a model, we need to make sure that the two models $\mathcal{T}_1$ and $\mathcal{T}_2$ agree on the common part of the signature. Formally, we assume that there exists a model $\mathcal{T}_0$ of the signature (S_0, Σ_0, Π_0) such that $\mathcal{T}_1\restriction_{h_1} = \mathcal{T}_0 = \mathcal{T}_2\restriction_{h_2}$ (i.e. the reduct of $\mathcal{T}_1$ through h_1 is the same as the reduct of $\mathcal{T}_2$ through h_2). Figure 6 below summarizes the construction of $\mathcal{T}$.

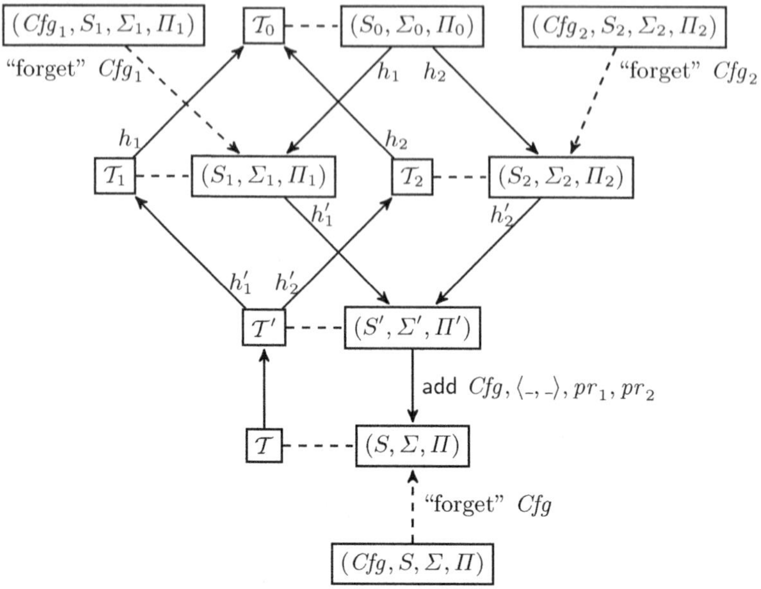

Fig. 6. Construction of amalgamated model

We can combine the two languages through the model amalgamation theorem. The theorem is not new (see for example [17]). A proof can also be found in our technical report [5].

Theorem 2 (Amalgamation). *If $\mathcal{T}_1$, $\mathcal{T}_2$ and $\mathcal{T}_0$ are models of (S_1, Σ_1, Π_1), (S_2, Σ_2, Π_2) and respectively (S_0, Σ_0, Π_0) such that $\mathcal{T}_1\restriction_{h_1} = \mathcal{T}_0 = \mathcal{T}_2\restriction_{h_2}$, there exists a unique model $\mathcal{T}'$ of (S', Σ', Π') such that $\mathcal{T}'\restriction_{h'_2} = \mathcal{T}_2$ and $\mathcal{T}'\restriction_{h'_1} = \mathcal{T}_1$.*

In order to obtain the model $\mathcal{T}$ for the aggregated signature $(\mathit{Cfg}, S, \Sigma, \Pi)$, we need to augment $\mathcal{T}'$ to interpret the Cfg sort, the pairing symbol $\langle \_, \_ \rangle$ and the projection symbols pr_1 and pr_2. Formally, we define $\mathcal{T}$ as follows:

1. $\mathcal{T}_{\mathit{Cfg}} = \mathcal{T}'_{\mathit{Cfg}'_1} \times \mathcal{T}'_{\mathit{Cfg}'_2}$,
2. $\mathcal{T}_{\langle \_, \_ \rangle}(\gamma_1, \gamma_2) = (\gamma_1, \gamma_2)$ for any $\gamma_1 \in \mathcal{T}'_{\mathit{Cfg}'_1}$ and any $\gamma_2 \in \mathcal{T}'_{\mathit{Cfg}'_2}$,
3. $\mathcal{T}_{pr_1}((\gamma_1, \gamma_2)) = \gamma_1$ for any $(\gamma_1, \gamma_2) \in \mathcal{T}_{\mathit{Cfg}} = \mathcal{T}'_{\mathit{Cfg}'_1} \times \mathcal{T}'_{\mathit{Cfg}'_2}$,
4. $\mathcal{T}_{pr_2}((\gamma_1, \gamma_2)) = \gamma_2$ for any $(\gamma_1, \gamma_2) \in \mathcal{T}_{\mathit{Cfg}} = \mathcal{T}'_{\mathit{Cfg}'_1} \times \mathcal{T}'_{\mathit{Cfg}'_2}$, and
5. $\mathcal{T}_o = \mathcal{T}'_o$ for any other object $o \in S \cup \Sigma \cup \Pi$.

The aggregated signature and model have the following properties:

Lemma 1. *For any basic patterns $\pi_1 \in T_{\mathit{Cfg}_1}(\mathit{Var})$ and $\pi_2 \in T_{\mathit{Cfg}_2}(\mathit{Var})$, for any aggregated configuration $(\gamma_1, \gamma_2) \in \mathcal{T}_{\mathit{Cfg}}$, for any (well-sorted) valuation ρ, we have that*

$$\mathcal{T}, (\gamma_1, \gamma_2), \rho \models \langle \pi_1, \pi_2 \rangle \textit{ iff } \mathcal{T}_1, \gamma_1, \rho\restriction_{h'_1(S_1)} \models \pi_1 \textit{ and } \mathcal{T}_2, \gamma_2, \rho\restriction_{h'_2(S_2)} \models \pi_2.$$

Proof. We show the direct implication first. Assume $\mathcal{T}, (\gamma_1, \gamma_2), \rho \models \langle \pi_1, \pi_2 \rangle$. By definition, we have that $\rho(\langle \pi_1, \pi_2 \rangle) = (\gamma_1, \gamma_2)$. By the interpretation of $\langle \_, \_ \rangle$ in the model $\mathcal{T}$, we have that $\rho\lceil_{h'_1(S_1)}(\pi_1) = \gamma_1$ and $\rho\lceil_{h'_2(S_2)}(\pi_2) = \gamma_2$. But this is equivalent to $\mathcal{T}_1, \gamma_1, \rho\lceil_{h'_1(S_1)} \models \pi_1$ and $\mathcal{T}_2, \gamma_2, \rho\lceil_{h'_2(S_2)} \models \pi_2$, which is what we had to show.

For the reverse direction, we assume that $\mathcal{T}_1, \gamma_1, \rho\lceil_{h'_1(S_1)} \models \pi_1$ and that $\mathcal{T}_2, \gamma_2, \rho\lceil_{h'_2(S_2)} \models \pi_2$. Therefore $\rho\lceil_{h'_1(S_1)}(\pi_1) = \pi_1$ and $\rho\lceil_{h'_2(S_2)}(\pi_2) = \pi_2$, which implies, by the interpretation of $\langle \_, \_ \rangle$ in $\mathcal{T}$, that $\rho(\langle \pi_1, \pi_2 \rangle) = (\gamma_1, \gamma_2)$. But this is equivalent to $\mathcal{T}, (\gamma_1, \gamma_2), \rho \models \langle \pi_1, \pi_2 \rangle$, which is what we had to prove.

5.3 Language Aggregation

Having shown how to construct the matching logic signature (Cfg, S, Σ, Π) and model $\mathcal{T}$ for the aggregated language, we now show how the transition relation $\longrightarrow$ for the aggregated language is defined and how to construct the axioms A of the aggregated language from the axioms A_1 and A_2 of the initial languages.

We identify three types of language aggregations, depending on how the $\longrightarrow$ transition relation is defined from $\longrightarrow_1$ and $\longrightarrow_2$. Each of the three constructions could be useful in various contexts:

1. $\longrightarrow_1 \otimes_a \longrightarrow_2$ is the asynchronous interleaving product of the two transition relations, i.e.
$$(\gamma_1, \gamma_2) \longrightarrow (\gamma'_1, \gamma'_2) \text{ if}$$
$$\gamma_1 = \gamma'_1 \text{ and } \gamma_2 \longrightarrow_2 \gamma'_2 \text{ or } \gamma_1 \longrightarrow_1 \gamma'_1 \text{ and } \gamma_2 = \gamma'_2,$$
2. $\longrightarrow_1 \otimes_p \longrightarrow_2$ is the parallel product of the two transition relations, i.e.
$$(\gamma_1, \gamma_2) \longrightarrow (\gamma'_1, \gamma'_2) \text{ if}$$
$$\gamma_1 \longrightarrow_1 \gamma'_1 \text{ and } \gamma_2 \longrightarrow_2 \gamma'_2,$$
3. finally, $\longrightarrow_1 \otimes \longrightarrow_2 = (\longrightarrow_1 \otimes_a \longrightarrow_2) \cup (\longrightarrow_1 \otimes_p \longrightarrow_2)$ is the (general) product of $\longrightarrow_1$ and $\longrightarrow_2$.

The asynchronous product with interleaving semantics means that in one step of the aggregated language, either the left-hand side takes a step (in the first language) or the right-hand side takes a step (in the second language). The parallel product forces both sides to take steps simultaneously. The (general) product requires at least one side to take a step and it allows (but not requires) the other side to do the same.

5.4 Constructing the Axioms for the Three Products

We next show how to construct a set of axioms A for the aggregated language from the set of axioms A_1 and A_2 of the initial languages, depending on which of the three constructions is chosen for the aggregated transition relation. The main result is that, for each of the three constructions, the transition relation is a model of the aggregated axioms. This means that we can construct the formal semantics of the aggregated language directly from the formal semantics of the initial languages.

Let A_1 be a set of pure reachability rules over the signature $(Cfg_1, S_1, \Sigma_1, \Pi_1)$ that capture the semantics of the first language (the axioms of the first language): $\mathcal{T}_1, \longrightarrow_1 \models A_1$. Let A_2 be a set of pure reachability rules over the signature $(Cfg_2, S_2, \Sigma_2, \Pi_2)$ that capture the semantics of the second language (the axioms of the second language): $\mathcal{T}_2, \longrightarrow_2 \models A_2$.

In order to define the axioms for the aggregated language, we need a way to transform reachability formulae from the two initial signatures into reachability formulae of the target signature. This is performed with the help of the following function:

Definition 10. *We define the function ι_x^i (for $i \in \{1,2\}$ and x a distinguished variable in Var of sort Cfg_i) that takes as input a matching logic formula over $(Cfg_1, S_1, \Sigma_1, \Pi_1)$ (respectively $(Cfg_2, S_2, \Sigma_2, \Pi_2)$) and changes all basic patterns π into $\langle \pi, x\rangle$ (respectively $\langle x, \pi\rangle$) in order to obtain a formula over (Cfg, S, Σ, Π):*

1. $\iota_x^1(\pi) = \langle h_1'(\pi), x\rangle$
2. $\iota_x^2(\pi) = \langle x, h_2'(\pi)\rangle$
3. $\iota_x^i(\varphi_1 \wedge \varphi_2) = \iota_x^i(\varphi_1) \wedge \iota_x^i(\varphi_2)$,
4. $\iota_x^i(\exists y.\varphi_1) = \exists y.\iota_x^i(\varphi_1)$,
5. $\iota_x^i(\neg\varphi_1) = \neg\iota_x^i(\varphi_1)$ *and*
6. $\iota_x^i(P(t_1, \ldots, t_n)) = P(h_i'(t_1), \ldots, h_i'(t_n))$.

Example 8. If φ_1 is the matching formula $\langle \texttt{while}\ (E)\ S, \sigma\rangle \wedge eval(\sigma, E) \neq_{Int} 0$ and φ_2 is $\langle \texttt{letrec f x = if}\ (I)\ \texttt{then}\ E_1\ \texttt{else}\ E_2\ \texttt{in f(x)}\rangle \wedge I <_{Int} 5$, then $\iota_x^i(\varphi_1)$ is $\langle\langle \texttt{while}\ (E)\ S, \sigma\rangle, x\rangle \wedge eval(\sigma, E) \neq_{Int} 0$ and $\iota_x^i(\varphi_2)$ is the aggregate formula $\langle x, \langle \texttt{letrec f x = if}\ (I)\ \texttt{then}\ E_1\ \texttt{else}\ E_2\ \texttt{in f(x)}\rangle\rangle \wedge I <_{Int} 5$.

Next, we show the link, in terms of matching logic formulae, between the aggregated model and the initial model.

Lemma 2. *Let φ_1 and φ_2 be matching logic formulae over $(Cfg_1, S_1, \Sigma_1, \Pi_1)$ and respectively $(Cfg_2, S_2, \Sigma_2, \Pi_2)$. For any aggregated configuration $(\gamma_1, \gamma_2) \in \mathcal{T}_{Cfg}$, for any (well-sorted) valuation ρ, if x is a fresh variable, we have that*

$$\mathcal{T}, (\gamma_1, \gamma_2), \rho \models \iota_x^i(\varphi_i) \text{ iff } \mathcal{T}_i, \gamma_i, \rho\restriction_{h_i'(S_i)} \models \varphi_i.$$

Proof. For simplicity, we assume that $i = 1$ (since the case with $i = 2$ is analogous). We prove the lemma by structural induction on φ_1. We only show the case of negation, the other cases being similar:

– if $\varphi_1 = \neg\varphi_1'$, then

$$\begin{aligned}
&\mathcal{T}, (\gamma_1, \gamma_2), \rho \models \iota_x^1(\varphi_1) \text{ iff}\\
&\mathcal{T}, (\gamma_1, \gamma_2), \rho \models \neg\iota_x^1(\varphi_1') \text{ iff}\\
&\mathcal{T}, (\gamma_1, \gamma_2), \rho \not\models \iota_x^1(\varphi_1') \text{ iff}\\
&\mathcal{T}_1, \gamma_1, \rho\restriction_{h_1'(S_1)} \not\models \varphi_1' \text{ iff}\\
&\mathcal{T}_1, \gamma_1, \rho\restriction_{h_1'(S_1)} \models \neg\varphi_1'
\end{aligned}$$

Lemma 3. *For any pure matching logic formulae* φ_1 *over* $(Cfg_1, S_1, \Sigma_1, \Pi_1)$ *and* φ_2 *over* $(Cfg_2, S_2, \Sigma_2, \Pi_2)$, *for any aggregated configuration* $(\gamma_1, \gamma_2) \in \mathcal{T}_{Cfg}$, *for any (well-sorted) valuation* ρ, *if* x *is a fresh variable, we have that*

$$\mathcal{T}, (\gamma_1, \gamma_2), \rho \models \iota^i_x(\varphi_i) \textit{ iff } \mathcal{T}_i, \gamma_i, \rho\lceil_{h'_i(S_i)} \models \varphi_i \textit{ and } \rho(x) = \gamma_{3-i}.$$

Proof. For simplicity, we assume that $i = 1$ (since the case with $i = 2$ is analogous). We prove the lemma by structural induction on φ_1, showing that

$$\mathcal{T}, (\gamma_1, \gamma_2), \rho \models \iota^1_x(\varphi_1) \text{ iff } \mathcal{T}_1, \gamma_1, \rho\lceil_{h'_1(S_1)} \models \varphi_1 \text{ and } \rho(x) = \gamma_2.$$

We distinguish the following cases:

1. if $\varphi_1 = \pi$, then $\iota^1_x(\varphi_1) = (h'_1(\pi), x)$. We have that

$$\begin{array}{ll} \mathcal{T}, (\gamma_1, \gamma_2), \rho \models \iota^1_x(\varphi_1) & \text{iff} \\ \mathcal{T}, (\gamma_1, \gamma_2), \rho \models (h'_1(\pi), x) & \text{iff} \\ (\gamma_1, \gamma_2) = \rho(h'_1(\pi), x) & \text{iff} \\ \gamma_1 = \rho(h'_1(\pi)) \text{ and } \gamma_2 = \rho(x) & \text{iff} \\ \gamma_1 = \rho\lceil_{h'_1(S_1)}(\pi) \text{ and } \gamma_2 = \rho(x) & \text{iff} \\ \mathcal{T}_1, \gamma_1, \rho\lceil_{h'_1(S_1)} \models \pi \text{ and } \gamma_2 = \rho(x). & \end{array}$$

2. if $\varphi_1 = \exists y.\varphi'_1$ for a variable y of sort s (with $y \neq x$ since x is a fresh variable) then $\iota^1_x(\varphi_1) = \exists y.\iota^1_x(\varphi'_1)$. We have that

$$\begin{array}{ll} \mathcal{T}, (\gamma_1, \gamma_2), \rho \models \iota^1_x(\varphi_1) & \text{iff} \\ \mathcal{T}, (\gamma_1, \gamma_2), \rho \models \exists y.\iota^1_x(\varphi'_1) & \text{iff} \\ \text{there is } u \in \mathcal{T}_{1s} \text{ s.t. } \mathcal{T}, (\gamma_1, \gamma_2), \rho[y \mapsto u] \models \iota^1_x(\varphi'_1) & \text{iff} \\ \text{there is } u \in \mathcal{T}_{1s} \text{ s.t. } \mathcal{T}_1, \gamma_1, \rho[y \mapsto u]\lceil_{h'_1(S_1)} \models \varphi'_1 \text{ and } \gamma_2 = \rho[y \mapsto u](x) & \text{iff} \\ \mathcal{T}_1, \gamma_1, \rho\lceil_{h'_1(S_1)} \models \exists y.\varphi'_1 \text{ and } \gamma_2 = \rho(x) & \text{iff} \\ \mathcal{T}_1, \gamma_1, \rho\lceil_{h'_1(S_1)} \models \varphi_1 \text{ and } \gamma_2 = \rho(x). & \end{array}$$

3. if $\varphi_1 = \varphi'_1 \wedge \varphi'_2$ for a pure formula φ'_1 and a patternless formula φ'_2, then $\iota^1_x(\varphi_1) = \iota^1_x(\varphi'_1) \wedge \iota^1_x(\varphi'_2)$. We have that

$$\begin{array}{ll} \mathcal{T}, (\gamma_1, \gamma_2), \rho \models \iota^1_x(\varphi_1) & \text{iff} \\ \mathcal{T}, (\gamma_1, \gamma_2), \rho \models \iota^1_x(\varphi'_1) \wedge \iota^1_x(\varphi'_2) & \text{iff} \\ \mathcal{T}, (\gamma_1, \gamma_2), \rho \models \iota^1_x(\varphi'_1) \text{ and } \mathcal{T}, (\gamma_1, \gamma_2), \rho \models \iota^1_x(\varphi'_2) & \text{iff} \\ \mathcal{T}_1, \gamma_1, \rho\lceil_{h'_1(S_1)} \models \varphi'_1 \text{ and } \rho(x) = \gamma_2 \text{ and } \mathcal{T}_1, \gamma_1, \rho\lceil_{h'_1(S_1)} \models \varphi'_2 & \text{iff} \\ \mathcal{T}_1, \gamma_1, \rho\lceil_{h'_1(S_1)} \models \varphi'_1 \wedge \varphi'_2 \text{ and } \rho(x) = \gamma_2 & \text{iff} \\ \mathcal{T}_1, \gamma_1, \rho\lceil_{h'_1(S_1)} \models \varphi_1 \text{ and } \rho(x) = \gamma_2. & \end{array}$$

In the following subsections, we define three types of aggregations for these sets of axioms, for each type of language aggregation.

The Axioms for the Asynchronous Product with Interleaving Semantics. We let

$$\begin{aligned} A_1 \otimes_a A_2 = {} & \{\iota^1_y(\varphi_1) \Rightarrow \iota^1_y(\varphi'_1) \mid \varphi_1 \Rightarrow \varphi'_1 \in A_1\} \cup \\ & \{\iota^2_x(\varphi_2) \Rightarrow \iota^2_x(\varphi'_2) \mid \varphi_2 \Rightarrow \varphi'_2 \in A_2\} \end{aligned}$$

where x is a fresh variable of sort Cfg'_1 and y is a fresh variable of sort Cfg'_2. The intuition is that x captures any left-hand side and allow the right-hand to take a step while y captures any right-hand side and allow the left-hand side to take a step. We show formally that $A_1 \otimes_a A_2$ is indeed a formal specification of the language $(\mathcal{T}, \longrightarrow_1 \otimes_a \longrightarrow_2)$:

Theorem 3 (correctness for asynchronous product). *Let $A = A_1 \otimes_a A_2$ and $\longrightarrow = \longrightarrow_1 \otimes_a \longrightarrow_2$. We have that*

$$(\mathcal{T}, \longrightarrow) \models A.$$

Proof. We have to show that $(\mathcal{T}, \longrightarrow) \models A$. By definition, $(\mathcal{T}, \longrightarrow) \models A$ if, for any reachability rule $\varphi \Rightarrow \varphi' \in A$, we have that $(\mathcal{T}, \longrightarrow) \models \varphi \Rightarrow \varphi'$. Let $\varphi \Rightarrow \varphi' \in A$ be an arbitrary reachability rule. We show that $(\mathcal{T}, \longrightarrow) \models \varphi \Rightarrow \varphi'$.

As $\varphi \Rightarrow \varphi' \in A = A_1 \otimes_a A_2$, it follows that $\varphi \Rightarrow \varphi' = \iota^1_y(\varphi_1) \Rightarrow \iota^1_y(\varphi'_1)$ for some $\varphi_1 \Rightarrow \varphi'_1 \in A_1$ and a fresh variable y or that $\varphi \Rightarrow \varphi' = \iota^2_x(\varphi_2) \Rightarrow \iota^2_x(\varphi'_2)$ for some $\varphi_2 \Rightarrow \varphi'_2 \in A_2$ and a fresh variable x. Since the two cases are analogous, we deal only with the first and we assume therefore that $\varphi \Rightarrow \varphi' = \iota^1_y(\varphi_1) \Rightarrow \iota^1_y(\varphi'_1)$ for some $\varphi_1 \Rightarrow \varphi'_1 \in A_1$ and a fresh variable y. Therefore it remains to show that $(\mathcal{T}, \longrightarrow) \models \iota^1_y(\varphi_1) \Rightarrow \iota^1_y(\varphi'_1)$.

Let ρ be an arbitrary valuation. Let $(\gamma_1, \gamma_2) \in \mathcal{T}_{Cfg}$ be an arbitrary configuration such that $\mathcal{T}, (\gamma_1, \gamma_2), \rho \models \iota^1_y(\varphi_1)$. By Lemma 3, we have $\mathcal{T}_1, \gamma_1, \rho\restriction_{h'_1(S_1)} \models \varphi_1$ and $\rho(x) = \gamma_2$. Given that $\varphi_1 \Rightarrow \varphi'_1 \in A_1$ and $\mathcal{T}_1, \longrightarrow_1 \models A_1$, there exists $\gamma'_1 \in \mathcal{T}_{1\,Cfg}$ such that $\gamma_1 \longrightarrow_1 \gamma'_1$ and $\mathcal{T}_1, \gamma'_1, \rho\restriction_{h'_1(S_1)} \models \varphi'_1$. By Lemma 3, we have that $\mathcal{T}, (\gamma'_1, \gamma_2), \rho \models \iota^1_y(\varphi'_1)$. But, by the definition of $\longrightarrow$ ($\longrightarrow = \longrightarrow_1 \otimes_a \longrightarrow_2$), we also have that $(\gamma_1, \gamma_2) \longrightarrow (\gamma'_1, \gamma_2)$.

We have shown that for an arbitrary valuation ρ and for an arbitrary configuration $(\gamma_1, \gamma_2) \in \mathcal{T}_{Cfg}$ such that $\mathcal{T}, (\gamma_1, \gamma_2), \rho \models \iota^1_y(\varphi_1)$, there exists a configuration (γ'_1, γ_2) such that $(\gamma_1, \gamma_2) \longrightarrow (\gamma'_1, \gamma_2)$ and $\mathcal{T}, (\gamma'_1, \gamma_2), \rho \models \iota^1_y(\varphi'_1)$. But this means that $(\mathcal{T}, \longrightarrow) \models \iota^1_y(\varphi_1) \Rightarrow \iota^1_y(\varphi'_1)$, which is what we had to show.

The Axioms for the Parallel Product. We let

$$A_1 \otimes_p A_2 = \{\iota^1_y(\varphi_1) \wedge \iota^2_x(\varphi_2) \Rightarrow \exists x.\exists y.(\iota^1_y(\varphi'_1) \wedge \iota^2_x(\varphi'_2)) \mid \\ \varphi_1 \Rightarrow \varphi'_1 \in A_1, \varphi_2 \Rightarrow \varphi'_2 \in A_2\}$$

where x is a fresh variable of sort Cfg'_1 and y is a fresh variable of sort Cfg'_2. The intuition is that x captures any left-hand side and y captures any right-hand side. We show formally that $A_1 \otimes_p A_2$ is indeed a formal specification of the language $(\mathcal{T}, \longrightarrow_1 \otimes_p \longrightarrow_2)$:

Theorem 4 (correctness for parallel product). *Let $A = A_1 \otimes_p A_2$ and $\longrightarrow = \longrightarrow_1 \otimes_p \longrightarrow_2$. We have that*

$$(\mathcal{T}, \longrightarrow) \models A.$$

Proof. We have to show that $(\mathcal{T}, \longrightarrow) \models A$. By definition, $(\mathcal{T}, \longrightarrow) \models A$ if, for any reachability rule $\varphi \Rightarrow \varphi' \in A$, we have that $(\mathcal{T}, \longrightarrow) \models \varphi \Rightarrow \varphi'$. Let $\varphi \Rightarrow \varphi' \in A$ be an arbitrary reachability rule. We show that $(\mathcal{T}, \longrightarrow) \models \varphi \Rightarrow \varphi'$.

As $\varphi \Rightarrow \varphi' \in A = A_1 \otimes_p A_2$, it follows that there exist $\varphi_1 \Rightarrow \varphi_1 \in A_1$, $\varphi_2 \Rightarrow \varphi_2' \in A_2$, fresh variables x and y such that $\varphi \Rightarrow \varphi' = \iota_y^1(\varphi_1) \wedge \iota_x^2(\varphi_2) \Rightarrow \iota_y^1(\varphi_1') \wedge \iota_x^2(\varphi_2')$.

Let ρ be an arbitrary valuation. Let (γ_1, γ_2) be an arbitrary configuration such that $\mathcal{T}, (\gamma_1, \gamma_2), \rho \models \iota_y^1(\varphi_1) \wedge \iota_x^2(\varphi_2)$. By the semantics of $\wedge$, we have that $\mathcal{T}, (\gamma_1, \gamma_2), \rho \models \iota_y^1(\varphi_1)$ and $\mathcal{T}, (\gamma_1, \gamma_2), \rho \models \iota_x^2(\varphi_2)$. By Lemma 3 it follows that $\mathcal{T}_1, \gamma_1, \rho\lceil_{h_1'(S_1)} \models \varphi_1$, $\rho(y) = \gamma_2$, $\mathcal{T}_2, \gamma_2, \rho\lceil_{h_2'(S_2)} \models \varphi_2$ and $\rho(x) = \gamma_1$.

Because $\mathcal{T}_1, \gamma_1, \rho\lceil_{h_1'(S_1)} \models \varphi_1$, $\varphi_1 \Rightarrow \varphi_1' \in A_1$ and $\mathcal{T}_1, \longrightarrow_1 \models A_1$, we obtain that there exists γ_1' such that $\gamma_1 \longrightarrow_1 \gamma_1'$ and $\mathcal{T}_1, \gamma_1', \rho\lceil_{h_1'(S_1)} \models \varphi_1'$. Similarly, there exists γ_2' such that $\gamma_2 \longrightarrow_2 \gamma_2'$ and $\mathcal{T}_2, \gamma_2', \rho\lceil_{h_2'(S_2)} \models \varphi_2'$.

By Lemma 3, it follows that $\mathcal{T}, (\gamma_1', \gamma_2'), \rho[x \mapsto \gamma_1][y \mapsto \gamma_2] \models \iota_y^1(\varphi_1')$ and $\mathcal{T}, (\gamma_1', \gamma_2'), \rho[x \mapsto \gamma_1][y \mapsto \gamma_2] \models \iota_y^2(\varphi_2')$, which implies that $\mathcal{T}, (\gamma_1', \gamma_2'), \rho \models \exists x.\exists y.(\iota_y^1(\varphi_1') \wedge \iota_y^2(\varphi_2'))$. By the definition of $\longrightarrow$, we also have $(\gamma_1, \gamma_2) \longrightarrow (\gamma_1', \gamma_2')$.

We have started with an arbitrary valuation ρ and an arbitrary configuration (γ_1, γ_2) such that $\mathcal{T}, (\gamma_1, \gamma_2), \rho \models \iota_y^1(\varphi_1) \wedge \iota_x^2(\varphi_2)$ and we have shown that there exists a configuration (γ_1', γ_2') such that $(\gamma_1, \gamma_2) \longrightarrow (\gamma_1', \gamma_2')$ and $\mathcal{T}, (\gamma_1', \gamma_2'), \rho \models \exists x.\exists y.\iota_y^1(\varphi_1') \wedge \iota_y^2(\varphi_2')$. Therefore $(\mathcal{T}, \longrightarrow) \models \iota_y^1(\varphi_1) \wedge \iota_x^2(\varphi_2) \Rightarrow \exists x.\exists y.\iota_y^1(\varphi_1') \wedge \iota_y^2(\varphi_2'$, which is what we had to show.

The Axioms for the General Product. For the general product, which allows for both interleaving and parallel steps, we define

$$A_1 \otimes A_2 = A_1 \otimes_a A_2 \cup A_1 \otimes_p A_2.$$

The correctness result for the general product follows quickly from Theorems 3 and 4.

Theorem 5 (correctness for the general product). *Let $A = A_1 \otimes A_2$ and $\longrightarrow = \longrightarrow_1 \otimes \longrightarrow_2$. We have that*

$$(\mathcal{T}, \longrightarrow) \models A.$$

Proof. Let $\varphi \Rightarrow \varphi' \in A$ be an arbitrary rule. We show that $(\mathcal{T}, \longrightarrow) \models \varphi \Rightarrow \varphi'$. Let $\gamma \in \mathcal{T}_{Cfg}$ be an arbitrary configuration and let ρ be an arbitrary valuation such that $\mathcal{T}, \gamma, \rho \models \varphi$.

We distinguish two cases:

1. if $\varphi \Rightarrow \varphi' \in A_1 \otimes_a A_2$, then, by Theorem 3, there exists γ' such that $\gamma(\longrightarrow_1 \otimes_a \longrightarrow_2)\gamma'$ and $\mathcal{T}, \gamma', \rho \models \varphi'$. But, by definition, $\longrightarrow_1 \otimes_a \longrightarrow_2 \subseteq \longrightarrow$. Therefore there exists γ' such that $\gamma \longrightarrow \gamma'$ and $\mathcal{T}, \gamma', \rho \models \varphi'$, which implies $(\mathcal{T}, \longrightarrow) \models \varphi \Rightarrow \varphi'$.
2. if $\varphi \Rightarrow \varphi' \in A_1 \otimes_p A_2$, then, by Theorem 4, there exists γ' such that $\gamma(\longrightarrow_1 \otimes_p \longrightarrow_2)\gamma'$ and $\mathcal{T}, \gamma', \rho \models \varphi'$. But, by definition, $\longrightarrow_1 \otimes_p \longrightarrow_2 \subseteq \longrightarrow$. Therefore there exists γ' such that $\gamma \longrightarrow \gamma'$ and $\mathcal{T}, \gamma', \rho \models \varphi'$, which implies $(\mathcal{T}, \longrightarrow) \models \varphi \Rightarrow \varphi'$.

6 Conclusion and Future Work

In this paper we have shown that if two programming languages are defined using reachability logic axioms and matching-logic based semantics, then we can effectively construct products of the two languages, such that a pair of programs belonging to the product can be executed asynchronously with interleaving, in parallel (synchronised), or a combination of the two. The construction can be automated in definitional frameworks like $\mathbb{K}$ [15].

A programming language definition consists of a signature and a semantics. In our approach, the signature is a many-sorted first-order signature, that includes both the syntax of the programming languages and the data structures required by the semantics. The category of the many sorted first-order signatures has colimits, in particular pushouts. Moreover, this category has the amalgamation property [9,16]. The semantics of the programming languages is given by transitions systems. The category of the transition systems has also several nice constructions [19]. We combine these constructions in order to get the definition for aggregated languages. The approach is flexible enough to allow various aggregations. The semantics of a programming language can be specified with (one-step) reachability logic formulae. We show that the specification of the aggregated language can be obtained from the specifications of the components.

We used many-sorted first-order signatures in order to make the presentation easier to follow. However, the syntax of programming languages is usually given by BNF rules, which correspond to order-sorted first-order signatures. Unfortunately, order-sorted first-order logic does not have pushouts of signatures and the amalgamation property. There are several approaches dealing with this issue, see, e.g., [1,10,17]. It is challenging to see which one of these is the best candidate for programming languages products and this will be investigated in the future. It is also interesting to see if the formalisation of the matching logic and reachability logic as institutions [2,3] could help.

An interesting observation can be made for the case where $\longrightarrow = \longrightarrow_1 \otimes \longrightarrow_2$. The transition system $(\mathcal{T}, \longrightarrow)$ is a product in the category of the transitions systems [19]. So, the syntax of the aggregation of the language is defined by a pushout (which is a shared sum) and the semantics by a product.

Language aggregation has uses in proving equivalence properties [4,6]. We intend to explore its use in proving other kinds of relations and in compiler verification.

References

1. Alpuente, M., Escobar, S., Meseguer, J., Ojeda, P.: Order-sorted generalization. Electr. Notes Theor. Comput. Sci. **246**, 27–38 (2009)
2. Chiriţă, C.E.: An institutional foundation for the k semantic framework. Master's thesis, University of Bucharest (2014)
3. Chiriţă, C.E., Şerbănuţă, T.F.: An institutional foundation for the k semantic framework. In: 22nd International Workshop Recent Trends in Algebraic Development Techniques, WADT 2014 (in press)

4. Ciobâcă, Ş.: Reducing partial equivalence to partial correctness. In: 2014 16th International Symposium on Symbolic and Numeric Algorithms for Scientific Computing (SYNASC), pp. 164–171, September 2014
5. Ciobâcă, Ş., Lucanu, D., Rusu, V., Roşu, G.: A language-independent proof system for mutual program equivalence. Technical report 14–01, Al. I. Cuza University (2014)
6. Ciobâcă, Ş., Lucanu, D., Rusu, V., Roşu, G.: A language-independent proof system for mutual program equivalence. In: Merz, S., Pang, J. (eds.) ICFEM 2014. LNCS, vol. 8829, pp. 75–90. Springer, Heidelberg (2014)
7. Clavel, M., Durán, F., Eker, S., Lincoln, P., Martí-Oliet, N., Meseguer, J., Talcott, C.L.: All About Maude, A High-Performance Logical Framework. LNCS, vol. 4350. Springer, Heidelberg (2007)
8. Ştefănescu, A., Ciobâcă, Ş., Mereuta, R., Moore, B.M., Şerbănută, T.F., Roşu, G.: All-path reachability logic. In: Dowek, G. (ed.) RTA-TLCA 2014. LNCS, vol. 8560, pp. 425–440. Springer, Heidelberg (2014)
9. Diaconescu, R.: Institution-Independent Model Theory. Birkhauser, Basel (2008)
10. Haxthausen, A.E., Nickl, F.: Pushouts of order-sorted algebraic specifications. In: Nivat, M., Wirsing, M. (eds.) AMAST 1996. LNCS, vol. 1101, pp. 132–147. Springer, Heidelberg (1996)
11. Roşu, G., Ellison, C., Schulte, W.: Matching logic: an alternative to Hoare/Floyd logic. In: Johnson, M., Pavlovic, D. (eds.) AMAST 2010. LNCS, vol. 6486, pp. 142–162. Springer, Heidelberg (2011)
12. Roşu, G.: Matching logic: a logic for structural reasoning. Technical report, University of Illinois, January 2014. http://hdl.handle.net/2142/47004
13. Roşu, G.: Matching logic (invited talk). In: 26th International Conference on Rewriting Techniques and Applications, RTA 2015, 29 June - 1 July, Warsaw, Poland (2015, to appear)
14. Roşu, G., Ştefanescu, A., Ştefan Ciobâcă, Moore, B.M.: One-path reachability logic. In: 28th Annual ACM/IEEE Symposium on Logic in Computer Science, LICS 2013, 25–28 June 2013, New Orleans, LA, USA, pp. 358–367 (2013). http://dx.doi.org/10.1109/LICS.2013.42
15. Roşu, G., Şerbănuţă, T.F.: An overview of the K semantic framework. J. Log. Algebr. Program. **79**(6), 397–434 (2010)
16. Sannella, D., Tarlecki, A.: Foundations of Algebraic Specification and Formal Software Development. Monographs in Theoretical Computer Science. An EATCS Series. Springer, Heidelberg (2012). http://dx.doi.org/10.1007/978-3-642-17336-3
17. Schröder, L., Mossakowski, T., Tarlecki, A., Hoffman, P., Klin, B.: Amalgamation in the semantics of CASL. Theoret. Comput. Sci. **331**(1), 215–247 (2005). http://dx.doi.org/10.1016/j.tcs.2004.09.037
18. Şerbănuţă, T.F., Roşu, G., Meseguer, J.: A rewriting logic approach to operational semantics. Inf. Comput. **207**(2), 305–340 (2009)
19. Winskel, G., Nielsen, M.: Categories in concurrency. In: Pitts, A.M., Dybjer, P. (eds.) Semantics and Logics of Computation, Publications of the Newton Institute, pp. 299–354. Cambridge University Press, Cambridge (1997)

Coalgebraic Semantics of Heavy-Weighted Automata

Marie Fortin[1,3], Marcello M. Bonsangue[2,3(✉)], and Jan Rutten[3,4]

[1] École Normale Supérieure de Cachan, Cachan, France
[2] LIACS – Leiden University, Leiden, The Netherlands
m.m.bonsangue@liacs.leidenuniv.nl
[3] Centrum Wiskunde & Informatica, Amsterdam, The Netherlands
[4] ICIS – Radboud University Nijmegen, Nijmegen, The Netherlands

Abstract. In this paper we study *heavy-weighted automata*, a generalization of weighted automata in which the weights of the transitions can be formal power series. As for ordinary weighted automata, the behaviour of heavy-weighted automata is expressed in terms of formal power series. We propose several equivalent definitions for their semantics, including a system of *behavioural differential equations* (following the approach of *coinductive calculus*), or an embedding into a coalgebra for the functor $S \times (-)^A$, for which the set of formal power series is a final coalgebra. Using techniques based on bisimulations and coinductive calculus, we study how ordinary weighted automata can be transformed into more compact heavy-weighted ones.

1 Introduction

Weighted automata are a generalization of non-deterministic automata in which each transition carries a weight [5]. This weight is an element of a semiring, representing, for example, the cost or probability of taking the transition. Weighted automata have many different areas of application. Recently, for example, they have been used to solve counting problems, first in [10] with a procedure called *coinductive counting*, then in [4] with the *counting automata methodology*.

Whereas non-deterministic automata either accept or reject a word, weighted automata associate with each word the cost of its execution. Their semantics is thus defined in terms of *weighted languages*, also called *formal power series*, which are functions mapping words to weights.

Formal power series form themselves a semiring, and thus can be used as weights of transitions, yielding what we call *heavy-weighted automata*. In [4], such automata are used to give a compact representation of some combinatorial problems. One of our motivation here is to extend the coalgebraic setting existing for ordinary weighted automata [1,10] to study equivalence between heavy-weighted automata. In particular, we are interested in producing a more compact representation of some well-shaped infinite weighted automata, generalizing the examples given in [4].

M. Codescu et al. (Eds.): WADT 2014, LNCS 9463, pp. 48–68, 2015.
DOI: 10.1007/978-3-319-28114-8_4

A second motivation for the introduction of heavy-weighted automata is to provide something similar to what generalized automata (where transitions are labeled by regular expressions) are to ordinary automata. In particular, we will see that Brzozowski-McCluskey's *state elimination method* [3,14] to compute the regular expression associated with a finite automaton also works for weighted automata. The method is not new, but is a nice application of our definition of heavy-weighted automata.

Though heavy-weighted automata can be seen as classic weighted automata over the semiring of weighted languages, their standard semantics as such (given in terms of power series over the semiring of power series) is not so interesting. Instead, we define the semantics of heavy-weighted automata in terms of (ordinary) power series, in three equivalent ways:

- by a system of equations linking the semantics of the different states.
- in terms of the final homomorphism. Here the set of weighted languages is the final coalgebra for the functor $S \times (-)^A$, which means that from any $S \times (-)^A$-coalgebra, there is a unique coalgebra homomorphism to the set of all weighted languages. Heavy-weighted automata are not themselves $S \times (-)^A$-coalgebras, but they can be embedded into one, using some kind of determinization procedure (as introduced in [13]).
- by giving a procedure that transforms a heavy-weighted automaton into an ordinary weighted automaton. This is done by composing ordinary weighted automata that recognize the weighted languages labeling the transitions of the heavy-weighted automaton.

We proceed as follows. First we briefly discuss some related work. Then, in Sect. 2 we recall the basic notions on formal power series, coinductive calculus and (ordinary) weighted automata. In Sect. 3, we define heavy-weighted automata and their behaviour, first in terms of behavioural differential equations and then in terms of a final homomorphism of coalgebras. In Sect. 4, we give a new interpretation to the behaviour of heavy-weighted automata, by giving a procedure that transforms a heavy-weighted automata into an ordinary weighted automata. In Sect. 5.1, we recall the state elimination method, and in Sect. 5.2, we show how heavy-weighted automata can be used to give a more compact representation for certain well-shaped infinite weighted automata.

Missing proof details can be found in the extended technical report [6].

Related Work

Our study of heavy-weighted automata is motivated by the work done in [4]. *Counting automata*, which are also automata having power series as weights, are used to model combinatorial problems. Some examples of reductions of infinite weighted automata to finite ones are given; yet there is no general format for such reductions. In Sect. 5.2 we describe a generalization of these reductions to a particular class of well-shaped, infinite weighted automata.

Automata in which transitions are labeled by power series were already present in [12]. The *state elimination method* (see Sect. 5.1) is also mentioned, though rather as the resolution of a system of equations. Apart from that, our work has no real intersection with what is done in [12]. A first difference is that we are interested in both finite and infinite atomata, whereas [12] focuses mostly on the finite case. Furthermore, no real separation is made in [12] between automata with ordinary weights and automata with power series as weights; as a consequence, the question of the transformation of a heavy-weighted automaton into an ordinary weighted automaton (see Sect. 4) is not raised.

Our definition of heavy-weighted automata present some differences with both [4,12]. In particular, we choose to define heavy-weighted automata in such a way that their behaviour is always defined, by requiring finite branching and not allowing ε-transitions. Thus we do not investigate the convergence issues that are given some importance in [4,12].

Our coalgebraic approach to heavy-weighted automata is new, and follows previous work on (ordinary) weighted automata, see e.g. [1,9,10]. Our construction in Sect. 3 can be seen as an instance of the generalized determinization construction described in [13]. Finally, our work relies largely on *coinductive calculus* for streams and power series, see e.g. [9,11].

2 Preliminaries

2.1 Coalgebras

We recall some basic definitions about coalgebras. Given a functor $\mathcal{F}$: Set $\to$ Set, an *$\mathcal{F}$-coalgebra* is a pair (X, f) consisting of a set X and a function $f : X \to \mathcal{F}X$.

An *$\mathcal{F}$-homomorphism* from an $\mathcal{F}$-coalgebra (X, f) to another $\mathcal{F}$-coalgebra (Y, g) is a function $h : X \to Y$ such that diagram to the right commutes, i.e. such that $g \circ h = \mathcal{F}h \circ f$. An $\mathcal{F}$-coalgebra (Y, g) is called *final* if for any $\mathcal{F}$-coalgebra (X, f) there exists a unique $\mathcal{F}$-homomorphism $[\![-]\!] : X \to Y$.

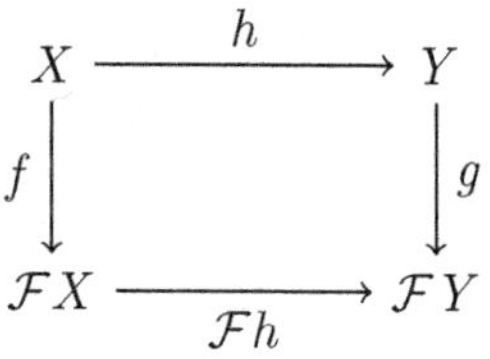

2.2 Formal Power Series

Throughout the paper, A denotes a nonempty finite alphabet.

Weighted automata associate to each input word a certain weight: their behaviour is defined in terms of *weighted languages*, also called *formal power series*, which are functions mapping words to elements of a semiring.

A *semiring* $(S, +, \times, 0, 1)$ consists of a set S together with two binary operations $+$ and $\times$ and two constants $0, 1 \in S$, such that:

(1) $(S, +, 0)$ is a commutative monoid and $(S, \times, 1)$ is a monoid;
(2) $\times$ distributes over $+$: $\forall x, y, z \in S,\ x \times (y + z) = x \times y + x \times z$ and $(x + y) \times z = x \times z + y \times z$;
(3) 0 is an annihilator for $\times$: $\forall x \in S, x \times 0 = 0 \times x = 0$.

Given a semiring S and a finite alphabet A, a *formal power series with coefficients in S and variables in A* is a function $\sigma : A^* \to S$. We denote by $S\langle\langle A\rangle\rangle$ the set of all formal power series with coefficients in S and variables in A.

Examples:

(1) Taking $S = \mathbb{B}$, where $\mathbb{B}$ denotes the boolean semiring, $\mathbb{B}\langle\langle A\rangle\rangle$ is isomorphic to the set of all formal languages over the alphabet A.
(2) $S\langle\langle \{X\}\rangle\rangle$ is isomorphic to the set of all streams with values in S (that is, the set of all functions $\mathbb{N} \to S$). We denote this set by S^ω, and we sometimes write $(s_0, s_1, \ldots)$ for the stream $i \mapsto s_i$.

The *support* of a formal power series σ is the set $\{w \in A^* \mid \sigma(w) \neq 0\}$. A *polynomial* is a power series with finite support. The set of all polynomials with coefficients in S and variables in A is denoted by $S\langle A\rangle$.

Finally, a power series $\sigma \in S\langle\langle A\rangle\rangle$ is called *proper* when $\sigma(\varepsilon) = 0$ (where ε is the empty word). We denote by $S\langle\langle A\rangle\rangle_p$ the set of all proper power series.

$S\langle\langle A\rangle\rangle$ can be given a $S \times (-)^A$-coalgebra structure using a generalization of the notion of *Brzozowski derivatives.* Given $a \in A$, the *a-derivative* $\sigma_a \in S\langle\langle A\rangle\rangle$ of a power series σ is defined by $\sigma_a(w) = \sigma(aw)$ and the *output* of σ is defined as $O(\sigma) = \sigma(\varepsilon)$. When A is a singleton, we write σ' for the derivative of σ.

We define $\Delta : S\langle\langle A\rangle\rangle \to S\langle\langle A\rangle\rangle^A$ by $\Delta(\sigma)(a) = \sigma_a$, and $\langle O, \Delta\rangle : S\langle\langle A\rangle\rangle \to S \times S\langle\langle A\rangle\rangle^A$ as the function $\sigma \mapsto (O(\sigma), \Delta(\sigma))$. Then $(S\langle\langle A\rangle\rangle, \langle O, \Delta\rangle)$ is a coalgebra for the functor $S \times (-)^A$. Moreover, we have the following theorem [9].

Theorem 1. *$(S\langle\langle A\rangle\rangle, \langle O, \Delta\rangle)$ is a final coalgebra for the functor $S \times (-)^A$.*

2.3 A Coinductive Calculus for Power Series

We now present some basic facts from the coinductive calculus for streams and power series developed in [9,11].

First we recall the coinduction proof principle, which will be one of our main proof techniques. A *bisimulation* on formal power series is a relation $\mathcal{R} \subseteq S\langle\langle A\rangle\rangle \times S\langle\langle A\rangle\rangle$ such that, for all σ and τ in $S\langle\langle A\rangle\rangle$, if $\sigma \mathrel{\mathcal{R}} \tau$ then

(1) $O(\sigma) = O(\tau)$;
(2) for all $a \in A$, $\sigma_a \mathrel{\mathcal{R}} \tau_a$.

The union of all bisimulation relations is called *bisimilarity*, and is denoted by $\sim$. A relation $\mathcal{R} \subseteq S\langle\langle A\rangle\rangle \times S\langle\langle A\rangle\rangle$ is a *bisimulation-up-to* if its closure under linear combination is a bisimulation relation [8].

Theorem 2 (Coinduction). *For all $\sigma, \tau \in S\langle\langle A\rangle\rangle$, if $\sigma \sim \tau$ then $\sigma = \tau$.*

Note that the converse trivially holds, since $\{(\sigma, \sigma) \mid \sigma \in S\langle\langle A\rangle\rangle\}$ is a bisimulation. The consequence of Theorem 2 is that to prove the equality of two power series σ and τ, it is sufficient to establish the existence of a bisimulation $\mathcal{R}$ such that $\sigma \mathrel{\mathcal{R}} \tau$.

Next, various operators on power series are defined coinductively. Coinductive definitions are given as *behavioural differential equations*, which have a unique solution. In particular, there exist a unique binary operator $+$, a unique binary operator $\times$, and for all $s \in S$ and $b \in A$, a unique $[s] \in S\langle\langle A\rangle\rangle$ and $[b] \in S\langle\langle A\rangle\rangle$ satisfying the following system of behavioural differential equations:

a-derivative (for all $a \in A$)	Initial value
$[s]_a = [0]$	$O([s]) = s$
$[b]_b = [1]$, $[b]_a = [0]$ for $a \neq b$	$O([b]) = 0$
$(\sigma + \tau)_a = \sigma_a + \tau_a$	$O(\sigma + \tau) = O(\sigma) + O(\tau)$
$(\sigma \times \tau)_a = (\sigma_a \times \tau) + ([O(\sigma)] \times \tau_a)$	$O(\sigma \times \tau) = O(\sigma) \times O(\tau)$

We then have for all $s \in S$, $[s](\varepsilon) = s$ and $[s](w) = 0$ if $w \neq \varepsilon$. For $b \in A$, $[b](b) = 1$ and $[b](w) = 0$ if $w \neq b$. The coinductive definitions given for the sum and convolution product coincide with the classic pointwise definitions: for all $\sigma, \tau \in S\langle\langle A\rangle\rangle$ and $w \in A^*$,

$$(\sigma + \tau)(w) = \sigma(w) + \tau(w) \quad \text{and} \quad (\sigma \times \tau)(w) = \sum_{uv=w} \sigma(u)\tau(v).$$

When S is a ring, we also define the inverse σ^{-1} of series σ such that $O(\sigma)$ is invertible in S, as the unique solution to $(\sigma^{-1})_a = -[O(\sigma)^{-1}] \times \sigma_a \times \sigma^{-1}$ and $O(\sigma^{-1}) = O(\sigma)^{-1}$. We then have $\sigma \times \sigma^{-1} = [1] = \sigma^{-1} \times \sigma$.

Theorem 3 (Fundamental Theorem). *For all $\sigma \in S\langle\langle A\rangle\rangle$,*

$$\sigma = [O(\sigma)] + \sum_{a\in A} [a] \times \sigma_a.$$

As a notational convenience, we will write s for $[s]$ and b for $[b]$ whenever it is clear from the context whether we intend elements of S and A or formal power series. Similarly, we will identify a word $w = a_1 \dots a_n \in A^*$ with the product $[a_1] \times \dots \times [a_n]$.

With these conventions, for all $\sigma \in S\langle\langle A\rangle\rangle$, $\sigma = \sum_{w\in A^*} \sigma(w) \times w$.

2.4 Rational Power Series

A family $\{\sigma_i \mid i \in I\}$ of power series is called *locally finite* when for all $w \in A^*$, the set $I_w = \{i \mid \sigma_i(w) \neq 0\}$ is finite. In this case, we define the sum $\sum_{i\in I} \sigma_i$ by $\left(\sum_{i\in I} \sigma_i\right)(w) = \sum_{i\in I_w} \sigma_i(w)$.

Let σ be a *proper* power series. For all $n \in \mathbb{N}$, we denote by σ^n the n-fold product of σ with itself: $\sigma^0 = 1$, and $\sigma^{n+1} = \sigma \times \sigma^n$. Then for all $w \in A^*$ and $n > |w|$, $\sigma^n(w) = 0$. Hence $\{\sigma^n \mid n \in \mathbb{N}\}$ is locally finite. We can thus define the *star* of a proper power series σ as the sum $\sigma^* = \sum_{n\in\mathbb{N}} \sigma^n$.

We define the set $\mathsf{RatE}_S(A)$ of all *rational S-expressions E* as follows:

$$E ::= s \in S \mid a \in A \mid (E + E) \mid (E \times E) \mid E^* .$$

We then define simultaneously the set of *valid* rational S-expressions, and the power series $\mathsf{val}(E)$ denoted by a valid expression, by induction:

- for all $s \in S$, s is valid and $\mathsf{val}(s) = s$;
- For all $a \in A$, a is valid and $\mathsf{val}(a) = a$;
- if E_1 and E_2 are valid, $(E_1 + E_2)$ is valid and $\mathsf{val}(E_1 + E_2) = \mathsf{val}(E_1) + \mathsf{val}(E_2)$;
- if E_1 and E_2 are valid, $(E_1 \times E_2)$ is valid and $\mathsf{val}(E_1 \times E_2) = \mathsf{val}(E_1) \times \mathsf{val}(E_2)$;
- if E is valid and $\mathsf{val}(E)$ is proper, E^* is valid and $\mathsf{val}(E^*) = \mathsf{val}(E)^*$.

A power series $\sigma \in S\langle\langle A\rangle\rangle$ is called *rational* if there exists a valid rational S-expression E such that $\mathsf{val}(E) = \sigma$. We denote by $S_{\mathrm{rat}}\langle\langle A\rangle\rangle$ the set of all rational power series.

2.5 Weighted Automata

Weighted automata are a generalisation of automata, where each transition has a weight in addition to the input letter. We associate a weight with each path in the automaton by multiplying the weights of all taken transitions; and we associate a weight with each word by adding the weights of all paths accepting it.

Let S be a semiring and A a finite alphabet. For any set X, we denote by $X \to_f S$ the set of all functions $g : X \to S$ such that $\{x \in X \mid g(x) \neq 0\}$ is finite.

Formally, a *weighted automaton* (or WA, for short) with input alphabet A and weights in the semiring S consists of a pair $(Q, \langle o, t\rangle)$, where:

- Q is a set of *states*.
- $o : Q \to S$ is the *output function*.
- $t : Q \to (Q \to_f S)^A$ is the *transition function*.

We will write $p \xrightarrow{a,s} q$ for $t(p)(a)(q) = s$, and $p \xrightarrow{s}$ for $o(p) = s$. A state $q \in Q$ is called *final* when $o(q) \neq 0$.

The *behaviour* $\mathcal{S}(q)$ of a state $q \in Q$, or weighted language recognized by state q, is classically defined as follows: for all $w = a_1 \dots a_n \in A^*$,

$$\mathcal{S}(q)(w) = \sum_{q_1,\dots,q_n \in Q} t(q)(a_1)(q_1) \times t(q_1)(a_2)(q_2) \times \cdots \times t(q_{n-1})(a_n)(q_n) \times o(q_n).$$

Note that this is a finite sum, since for all $q_i \in Q$ there are only finitely many $q_{i+1} \in Q$ such that $t(q_i)(a_{i+1})(q_{i+1}) \neq 0$. This follows the intuitive definition we gave before: an accepting path for w starting at state q is of the form $q = q_0 \xrightarrow{a_1,s_1} q_1 \xrightarrow{a_2,s_2} \cdots \xrightarrow{a_n,s_n} q_n \xrightarrow{s}$, with $s_i \neq 0$ and $s \neq 0$. Taking the sum of $s_1 \dots s_n s$ for all such paths, we obtain the given expression. However, we will mostly use the following equivalent coinductive definition: $\mathcal{S} : Q \to S\langle\langle A\rangle\rangle$ is defined as the unique solution to the following system of behavioural differential equations: for all $q \in Q$ and $a \in A$,

$$\mathcal{S}(q)_a = \sum_{r \in Q} t(q)(a)(r) \times \mathcal{S}(r) \qquad\qquad O(\mathcal{S}(q)) = o(q)\,.$$

Using the fundamental theorem of coinductive calculus, this is equivalent to

$$\mathcal{S}(q) = o(q) + \sum_{r \in Q} \left(\sum_{a \in A} a \times t(q)(a)(r) \right) \times \mathcal{S}(r)\,.$$

Example. Take $A = \{X\}$, $S = (\mathbb{R}, +, \times)$ and the automaton defined to the right. We have $\mathcal{S}(q_0) = X \times \mathcal{S}(q_1)$ and $\mathcal{S}(q_1) = 1 + X \times \mathcal{S}(q_0) + X \times \mathcal{S}(q_1)$. Recall that, as defined in Sect. 2.3, X is the stream $(0, 1, 0, 0, \ldots)$. These equations lead to $S(q_0) = X \times (1 - X - X^2)^{-1} = (0, 1, 1, 2, 3, \ldots)$, which corresponds to the Fibonacci sequence.

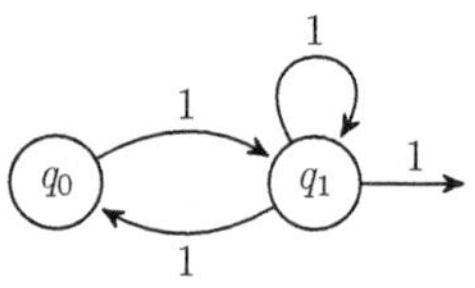

A weighted automaton is called *finite* when its set of states is finite. A power series $\sigma \in S\langle\langle A\rangle\rangle$ is *recognizable* when there exist a finite weighted automaton $\mathcal{A} = (Q, \langle o, t\rangle)$ and $q_0 \in Q$ such that $\mathcal{S}(q_0) = \sigma$. We denote by $S_{\text{rec}}\langle\langle A\rangle\rangle$ the set of all recognizable power series.

Theorem 4 (Kleene-Schutzenberger). $S_{rat}\langle\langle A\rangle\rangle = S_{rec}\langle\langle A\rangle\rangle$.

Note that when we don't require the set of states to be finite, for any power series $\sigma \in S\langle\langle A\rangle\rangle$, there exist a weighted automaton $(Q, \langle o, t\rangle)$ and $q_0 \in Q$ such that $\mathcal{S}(q_0) = \sigma$. For instance, take $Q = A^*$, $o = \sigma$, $t(w)(a)(aw) = 1$, and $t(w)(a)(v) = 0$ in all other cases. Then $\mathcal{S}(\varepsilon) = \sigma$.

3 Heavy-Weighted Automata

We generalize weighted automata to *heavy-weighted automata*, by allowing the weights of the transitions to be any power series rather than an element of the semiring S.

name	output function	transition function
Weighted automaton (WA)	$o : Q \to S$	$t : Q \to (Q \to_f S)^A$
Heavy-weighted automaton (HWA)	$o : Q \to S$	$t : Q \to (Q \to_f S\langle\langle A\rangle\rangle)^A$

Fig. 1. Definitions of WAs and HWAs

A *heavy-weighted automaton* (or HWA, for short) over the semiring S and the alphabet A consists of a pair $(Q, \langle o, t\rangle)$, where:

- Q is a set of *states*.
- $o : Q \to S$ is the *output function*.
- $t : Q \to (Q \to_f S\langle\langle A\rangle\rangle)^A$ is the *transition function*.

For any $p, q \in Q$, we also define the *cumulated weight between p and q* as

$$w(p)(q) = \sum_{a \in A} a \times t(p)(a)(q).$$

Note that for any $p, q \in Q$, $a \in A$, $w(p)(q)$ is proper and $t(p)(a)(q) = w(p)(q)_a$. As for ordinary weighted automata, we write $p \xrightarrow{a,\sigma} q$ for $t(p)(a)(q) = \sigma$, or $p \xrightarrow{\tau} q$ for $w(p)(q) = \tau$.

Remark. The transition function t is uniquely determined by w. Thus a HWA can equivalently be defined by giving its set of states Q, its output function $o : Q \to S$, and its cumulated weights, that is, a function $w : Q \to (Q \to_f S\langle\langle A\rangle\rangle_p)$. (Recall that here $S\langle\langle A\rangle\rangle_p$ denotes the set of all proper series.) Indeed, given any $w : Q \to (Q \to_f S\langle\langle A\rangle\rangle_p)$, define $t : Q \to (Q \to_f S\langle\langle A\rangle\rangle)^A$ by $t(p)(a)(q) = w(p)(q)_a$. The fundamental theorem then gives $w(p)(q) = \sum_{a \in A} a \times t(p)(a)(q)$.

Let $\mathcal{A} = (Q, \langle o, t\rangle)$ a HWA. The *behaviour* of a state $q \in Q$ is defined as a power series $\mathcal{S}(q) \in S\langle\langle A\rangle\rangle$, and satisfies the same equations as we had for WAs. More precisely, $\mathcal{S} : Q \to S\langle\langle A\rangle\rangle$ is defined as the unique solution to the following system of behavioural differential equations: for all $q \in Q$ and $a \in A$,

$$\mathcal{S}(q)_a = \sum_{r \in Q} t(q)(a)(r) \times \mathcal{S}(r) \qquad\qquad O(\mathcal{S}(q)) = o(q)\,.$$

HWAs are indeed a generalization of WAs, in the sense that any WA can be seen as a HWA, by identifying the weights in S with power series in $S\langle\langle A\rangle\rangle$. Since the behaviour of WAs and HWAs are defined by the same system of equations, the behaviour of a WA is unchanged when we consider it as a HWA.

Final Semantics for Heavy-Weighted Automata

In coalgebra theory, the behaviour of a system is usually defined in terms of final homorphism: given a functor $\mathcal{F}$ with a final coalgebra (Ω, ω), every element of an $\mathcal{F}$-coalgebra (X, f) is associated to a canonical representative in Ω by the final $\mathcal{F}$-homomorphism $[\![-]\!] : X \to \Omega$.

Here however, HWAs are coalgebras for the functor $X \mapsto S \times (X \to_f S\langle\langle A\rangle\rangle)^A$, whereas their semantics is defined in terms of formal power series, which is the final coalgebra for the functor $X \mapsto S \times X^A$.

The objective of this subsection is to propose another definition for the semantics of HWAs, equivalent to the previous one, but expressed in terms of final homomorphisms. For that, we will associate each HWA $(Q, \langle o, t\rangle)$ to an $S \times (-)^A$-coalgebra, in a construction similar to the determinization procedure for automata.

For the remainder of this subsection, we fix a HWA $\mathcal{A} = (Q, \langle o, t\rangle)$. Similarly to the powerset construction, we define a map $\eta : Q \to (Q \to_f S\langle\langle A\rangle\rangle)$ by

$$\eta(p)(q) = \begin{cases} 1 & \text{if } p = q \\ 0 & \text{if } p \neq q. \end{cases}$$

Note that every $\alpha : Q \to_f S\langle\langle A\rangle\rangle$ can be expressed as $\alpha = \sum_{q \in Q} \alpha(q) \cdot \eta(q)$ where, for all $\sigma \in S\langle\langle A\rangle\rangle$ and $\beta : Q \to_f S\langle\langle A\rangle\rangle$, $\sigma \cdot \beta : Q \to_f S\langle\langle A\rangle\rangle$ is defined as $q \mapsto \sigma \times \beta(q)$.

We want to define an $S \times (-)^A$ coalgebra structure $\langle \hat{o}, \hat{t}\rangle$ for $(Q \to_f S\langle\langle A\rangle\rangle)$ compatible with $\langle o, t\rangle$, meaning that $\hat{o} \circ \eta = o$ and $\hat{t} \circ \eta = t$. Moreover, $\hat{o}$ and $\hat{t}$ should behave as output and derivative functions, that is, the following equalities should hold for every $\alpha, \beta \in Q \to_f S\langle\langle A\rangle\rangle$, $\sigma \in S\langle\langle A\rangle\rangle$ and $a \in A$:

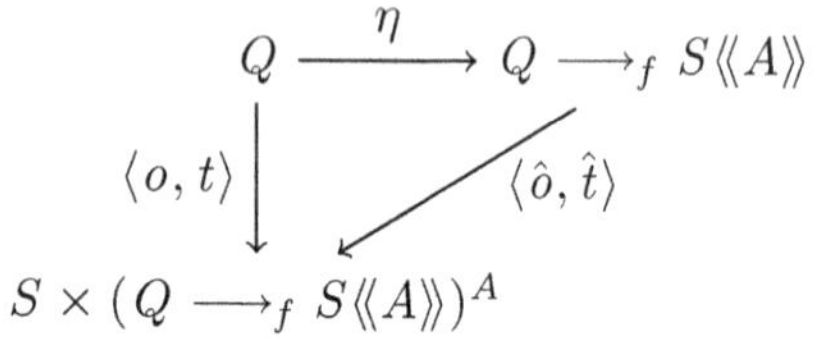

$$\hat{o}(\alpha + \beta) = \hat{o}(\alpha) + \hat{o}(\beta) \qquad \hat{t}(\alpha + \beta) = \hat{t}(\alpha) + \hat{t}(\beta)$$
$$\hat{o}(\sigma \cdot \alpha) = O(\sigma) \cdot \hat{o}(\alpha) \qquad \hat{t}(\sigma \cdot \alpha)(a) = \sigma_a \cdot \alpha + O(\sigma) \cdot \hat{t}(\alpha)$$

This leads to the following definitions for $\hat{o}$ and $\hat{t}$:

$$\hat{o}(\alpha) = \sum_{q \in Q} O(\alpha(q)) \cdot o(q) \qquad \hat{t}(\alpha)(a) = \sum_{q \in Q} (\alpha(q)_a \cdot \eta(q) + O(\alpha(q)) \cdot t(q)(a))$$

We are now going to exploit the fact that $S\langle\langle A\rangle\rangle$ is a final $S \times (-)^A$-coalgebra to define the semantics of $\mathcal{A}$. Denote by $[\![-]\!]$ the unique $S \times (-)^A$-homomorphism from $(Q \to_f S\langle\langle A\rangle\rangle)$ to $S\langle\langle A\rangle\rangle$.

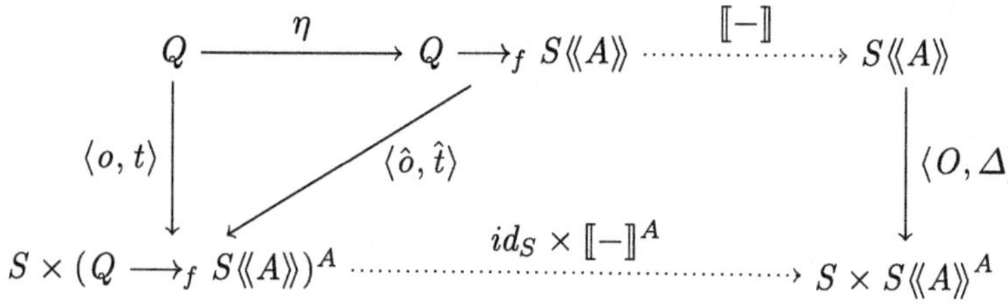

We now define the *behaviour* of state $q \in Q$ as the power series $[\![\eta(q)]\!]$, and show that it is indeed the same as the behaviour $\mathcal{S}(q)$ defined in Fig. 1.

Lemma 1. *For all $\alpha, \beta \in (Q \to_f S\langle\langle A\rangle\rangle)$ and $\sigma \in S\langle\langle A\rangle\rangle$,*

$$[\![\alpha + \beta]\!] = [\![\alpha]\!] + [\![\beta]\!] \quad \textit{and} \quad [\![\sigma \cdot \alpha]\!] = \sigma \times [\![\alpha]\!].$$

Proof. It is enough to show that $\mathcal{R}_1 = \{([\![\alpha + \beta]\!], [\![\alpha]\!] + [\![\beta]\!]) \mid \alpha, \beta : Q \to_f S\langle\langle A\rangle\rangle\}$ and $\mathcal{R}_2 = \{([\![\sigma \cdot \alpha]\!], \sigma \times [\![\alpha]\!]) \mid \sigma \in S\langle\langle A\rangle\rangle, \alpha : Q \to_f S\langle\langle A\rangle\rangle\}$ are a bisimulation and a bisimulation-up-to, using the fact that $O([\![\alpha]\!]) = \hat{o}(\alpha)$ and $[\![\alpha]\!]_a = [\![\hat{t}(\alpha)(a)]\!]$. □

Proposition 1. *For all $q \in Q$, $[\![\eta(q)]\!] = \mathcal{S}(q)$.*

Proof. We show that $q \mapsto [\![\eta(q)]\!]$ satisfies the system of equations defining $\mathcal{S}$:

- for all $q \in Q$, $O([\![\eta(q)]\!]) = \hat{o}(\eta(q)) = o(q)$.
- for all $q \in Q$ and $a \in A$,

$$[\![\eta(q)]\!]_a = [\![\hat{t}(\eta(q))(a)]\!] = [\![t(q)(a)]\!] = \left[\!\!\left[\sum_{r \in Q} t(q)(a)(r) \cdot \eta(r)\right]\!\!\right] = \sum_{r \in Q} t(q)(a)(r) \times [\![\eta(r)]\!]. \qquad \square$$

4 From Heavy-Weighted Automata to Weighted Automata

As we saw in Sect. 3, there is a trivial injection from the set of WAs to the set of HWAs. Reciprocally, we want to be able to transform any HWA $\mathcal{A} = (Q, \langle o, t \rangle)$ into a WA $\hat{\mathcal{A}} = \left(\hat{Q}, \langle \hat{o}, \hat{t} \rangle\right)$ with weights in S and input alphabet A, such that for all $q \in Q$, there exists $\hat{q} \in \hat{Q}$ such that q and $\hat{q}$ have the same behaviour. (Intuitively, this means that whatever state q we choose as "initial" in $\mathcal{A}$, we can find a state $\hat{q}$ in $\hat{\mathcal{A}}$ having the same behaviour.) There are two motivations for this construction:

- giving a new interpretation to the semantics of HWAs.
- giving a constructive proof that HWAs and WAs have the same expressivity. In the general case this is trivial since any power series can be recognized both by a HWA, and a (possibly infinite) WA, as shown in Sect. 2.5. Yet there are other interesting cases; for instance we can require the set of states to be finite, and the "heavy-weights" to be rational power series.

Let us first look at an example. We take $A = \{X\}$ and $S = \mathbb{R}$, and consider the automaton $\mathcal{A}_0$ on the right:

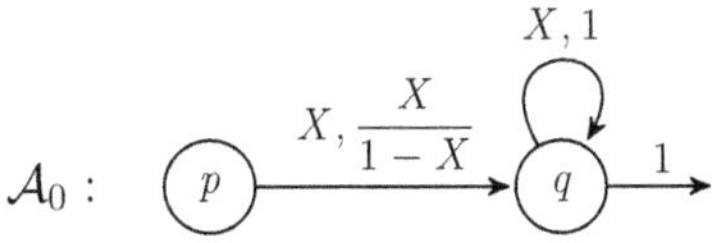

First Idea. We compute directly an equivalent WA by the method of "splitting the derivatives" [11]: we compute the successive derivatives of $\mathcal{S}(p)$, and add corresponding states at each step.

We have $\mathcal{S}(p)' = \frac{X}{1-X}\mathcal{S}(q)$, so we add a state with behaviour $\frac{X}{1-X}\mathcal{S}(q)$. Then $\left(\frac{X}{1-X}\mathcal{S}(q)\right)' = \frac{1}{1-X}\mathcal{S}(q)$, and we add a state with behaviour $\frac{1}{1-X}\mathcal{S}(q)$. Finally, $\left(\frac{1}{1-X}\mathcal{S}(q)\right)' = \frac{1}{1-X}\mathcal{S}(q) + \mathcal{S}(q)$, so we get the following automaton:

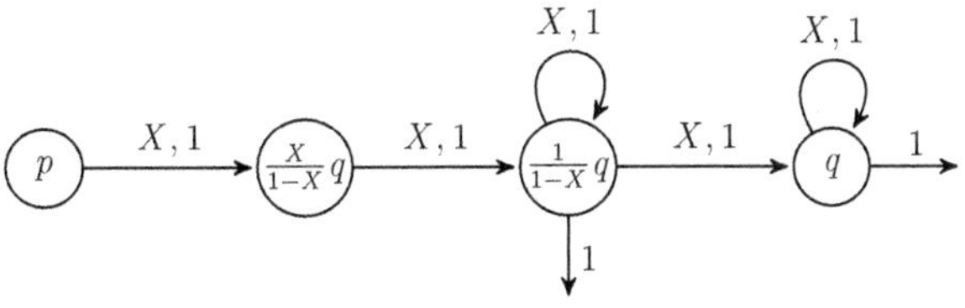

Second Idea. In the automaton on the right, the leftmost state recognize the stream $\frac{X}{1-X}$. We can plug it into $\mathcal{A}_0$, in place of the heavy-weighted transition, as follows:

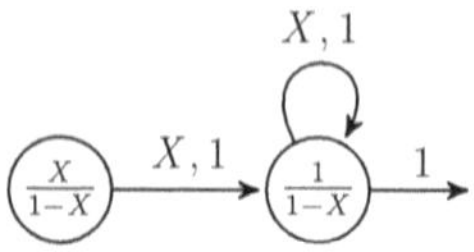

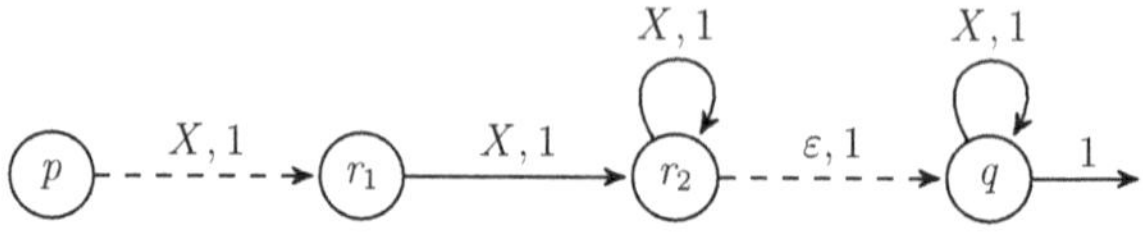

We did not allow ε-transitions in our definition of weighted automata, because the behaviour of a weighted automaton with ε-transitions is not always well-defined: for instance if we have a cycle of ε-transitions, we could have infinitely many paths labeled by the same word, which would lead to an infinite sum when computing the behaviour of a state. Yet here we don't add any infinite path labeled by ε, and in this particular case it is easy to adapt all the definitions.

Removing the ε-transitions, we get precisely the same automaton as with the first method. This is not surprising, since in the first method, computing the derivative of $\mathcal{S}(q)$ amounted to computing the derivative of $\frac{X}{1-X}$.

We come back to the general case. Both methods can be generalized, and lead again to the same definition of the equivalent WA.

Let $\mathcal{A} = (Q, \langle o, t \rangle)$ be a HWA. Assume that for all $p, q \in Q$ and $a \in A$, we have a WA $\mathcal{A}_{p,a,q} = (Q_{p,a,q}, \langle o_{p,a,q}, t_{p,a,q} \rangle)$ and a state $i_{p,a,q} \in Q_{p,a,q}$ with behaviour $t(p)(a)(q)$.

We define a WA $\hat{\mathcal{A}} = \left(\hat{Q}, \langle \hat{o}, \hat{t} \rangle\right)$ by setting:

- $\hat{Q} = Q \uplus \biguplus_{\substack{p,q \in Q \\ a \in A}} Q_{p,a,q}$
- $\forall q \in Q,\ \hat{o}(q) = o(q)$, and $\forall p, q \in Q, a \in A, r \in Q_{p,a,q},\ \hat{o}(r) = o_{p,a,q}(r) o(q)$
- $\forall p, q \in Q, a \in A$,

$$\hat{t}(p)(a)(i_{p,a,q}) = \begin{cases} 1 & \text{if } t(p)(a)(q) \neq 0 \\ 0 & \text{otherwise} \end{cases}$$

$\forall p, q \in Q, a, b \in A, r, s \in Q_{p,b,q}$,

$$\hat{t}(r)(a)(s) = \begin{cases} t_{p,b,q}(r)(a)(s) + o_{p,b,q}(r) & \text{if } p = q, a = b \text{ and } s = i_{p,a,p} \\ t_{p,b,q}(r)(a)(s) & \text{otherwise} \end{cases}$$

$\forall a, b \in A, p, q, q' \in Q$ s.t. $(p, b, q) \neq (q, a, q'), \forall r \in Q_{p,b,q}$,

$$\hat{t}(r)(a)(i_{q,a,q'}) = o_{p,b,q}(r)$$

In all other cases, $\hat{t}(r)(a)(s) = 0$.

This construction corresponds to the intuition we gave before (expressed as in the "second idea", though both methods lead to the same automaton). In fact, consider a transition $p \xrightarrow{b,\sigma} q$ in $\mathcal{A}$. We replace it by connecting $\mathcal{A}_{p,b,q}$

between p and q as we did in the example, using ε-transitions: we set a transition $p \xrightarrow{b,1} i_{p,b,q}$ and for each final state $r \in Q_{p,b,q}$ we set a transition $r \xrightarrow{\varepsilon, o_{p,b,q}(r)} q$, and set the output of r to 0.

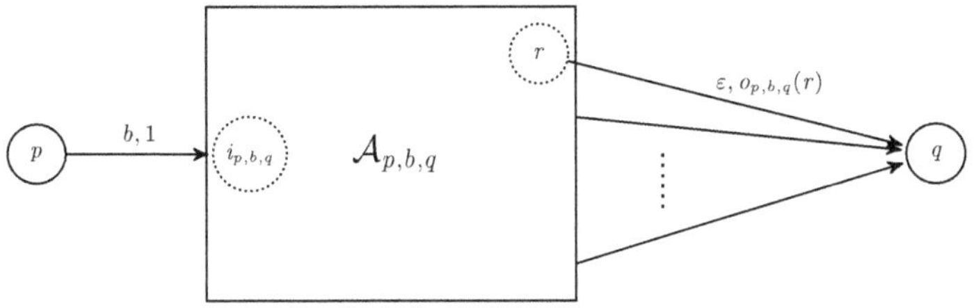

(There are no paths labeled by ε of length > 1, hence the semantics of the new automaton is well-defined.) We can then proceed to the removal of the ε-transitions : for every transition $r \xrightarrow{\varepsilon, o_{p,b,q}(r)} q$, we do the following : we remove the transition, we set $o(r) := o(r) + o_{p,b,q}(r)o(q)$, and for every $q \xrightarrow{a,s} q'$ we add a transition $r \xrightarrow{a, o_{p,b,q}(r)\cdot s} q'$. Note that the only such transitions $q \xrightarrow{a,s} q'$ are transitions of the form $q \xrightarrow{a,1} i_{q,a,q'}$.

Theorem 5. *With the above notations, denote by $\mathcal{S}$ the semantics of the automaton $\mathcal{A}$, and by $\hat{\mathcal{S}}$ the semantics of $\hat{\mathcal{A}}$. Then for all $q \in Q, \mathcal{S}(q) = \hat{\mathcal{S}}(q)$.*

Proof. It is enough to show that

$$\mathcal{R} = \left\{ \left(\hat{\mathcal{S}}(q), \mathcal{S}(q) \right) \mid q \in Q \right\} \cup \left\{ \left(\hat{\mathcal{S}}(r), \mathcal{S}_{p,a,q}(r)\mathcal{S}(q) \right) \mid p, q \in Q, r \in Q_{p,a,q} \right\}$$

is a bisimulation-up-to. (For all p, q, $\mathcal{S}_{p,a,q}$ denotes the semantics of $\mathcal{A}_{p,a,q}$.) □

Remark. Theorem 5 holds without any restriction on $\mathcal{A}$. Now consider the case where $\mathcal{A}$ is finite, and for all $p, q \in Q$, $t(p)(a)(q)$ is a rational power series. Then we can suppose that all $\mathcal{A}_{p,a,q}$ are also finite, and we obtain for $\hat{\mathcal{A}}$ a finite automaton as well. In particular, for all q, $\hat{\mathcal{S}}(q)$ is rational, i.e. $\mathcal{S}(q)$ is rational. This gives us a proof that (not surprisingly) finite HWAs with rational weights have the same expressivity as WAs. Yet there are other ways to prove this, for instance by directly computing $\mathcal{S}(q)$ as in Sect. 5.1.

5 Some Applications of Heavy-Weighted Automata

Heavy-weighted automata provide a more compact way of representing power series than ordinary weighted automata. We give two examples of how this can be used. First, there is the *state elimination method*, that describes a way to remove a state in a weighted automaton. In the case of finite automata, it also leads to an algorithm to compute a rational expression for the power series recognised by some state of the automaton. Secondly, we consider the case of infinite weighted automata representing algebraic power series. Under precise conditions on the shape of the automaton, we can formulate some contraction rules that lead to an equivalent, possibly finite, HWA.

5.1 State Elimination Method

Brzozowski and McCluskey's state elimination method for computing the rational expression associated to an (ordinary) finite automaton can easily be adapted to weighted automata. The only thing new with weighted automata is that we need to update also the outputs of the remaining states when we remove a state.

For practical reasons, we adopt in this subsection a slightly different definition of HWAs than in the rest of the paper. We now allow not only the weight of the transitions, but also the outputs of the states to be power series. Furthermore, we choose to define HWAs by giving their cumulated weight function w rather than t (see Sect. 3). Formally, a *heavy-weighted automaton* now is a pair $(Q, \langle o, w\rangle)$, where Q is a set of states, $o : Q \to S\langle\langle A\rangle\rangle$ is the output function, and $w : Q \to (Q \to_f S\langle\langle A\rangle\rangle_P)$.

The *behaviours* $\mathcal{S}(q)$ of each state $q \in Q$ are again defined as the unique solutions of a system of equations: for all $q \in Q$,

$$\mathcal{S}(q) = o(q) + \sum_{r \in Q} w(p)(q) \times \mathcal{S}(r).$$

Note that such an automaton can always be transformed into an automaton in which the output of all states are elements of S: we add one state f, with no outgoing transitions and output 1. For each other state $q \in Q$, we decompose $o(q)$ into $o(q) = s + \sigma$, with $s = O(o(q))$ and $\sigma = \sum_{a \in A} a \times o(q)_a$. Then we replace the ouptut of q by s, and we add a transition $q \xrightarrow{\sigma} f$. (Fig. 2).

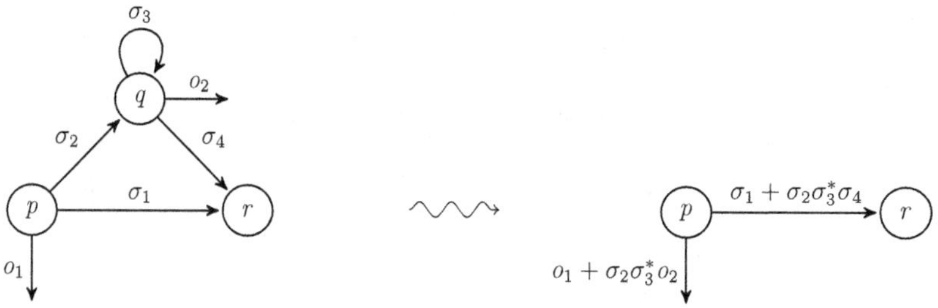

Fig. 2. Elimination of state q

State Elimination. Let $\mathcal{A} = (Q, \langle o, w\rangle)$ be a HWA with at least two states, and $q \in \mathcal{A}$. We define the automaton $\mathsf{elimination}(q, \mathcal{A})$ resulting from the elimination of state q in $\mathcal{A}$ as $\mathsf{elimination}(q, \mathcal{A}) = (Q \setminus \{q\}, \langle \hat{o}, \hat{w}\rangle)$, with for all $p, r \in Q$,

$$\hat{o}(p) = o(p) + w(p)(q) \times w(q)(q)^* \times o(q)$$
$$\hat{w}(p)(r) = w(p)(r) + w(p)(q) \times w(q)(q)^* \times w(q)(r).$$

We denote by $\mathcal{S}(p)$ the behaviour of a state $p \in Q$ in the automaton $\mathcal{A}$, and for $p \neq q$, we denote by $\hat{\mathcal{S}}(p)$ its behaviour in the automaton $\mathsf{elimination}(q, \mathcal{A})$.

Proposition 2. *For all $p \in Q \setminus \{q\}$, $\hat{\mathcal{S}}(p) = \mathcal{S}(p)$.*

Proof. $\mathcal{S}$ is defined by the following system of equations:

$$\forall p \in Q \qquad \mathcal{S}(p) = o(p) + \sum_{r \in Q} w(p)(r) \times \mathcal{S}(r)$$

The equation for q is equivalent to

$$\mathcal{S}(q) = w(q)(q)^* \Big(o(q) + \sum_{r \in Q \setminus \{q\}} w(q)(r)\mathcal{S}(r) \Big).$$

Substituting into the other equations, we get exactly the system defining $\hat{\mathcal{S}}$. □

Computing Rational Expressions for Finite Weighted Automata. The previous result holds without further restriction on $\mathcal{A}$. We now consider the case where Q is finite, and where all weights and outputs in $\mathcal{A}$ are given by rational expressions. Clearly, the weights and outputs in $\mathsf{eliminate}(q, \mathcal{A})$ can again be given by rational expressions, so these assumptions are preserved at each elimination of a state. In what follows, we identify rational expressions and the power series they denote.

Suppose we start from a finite ordinary weighted automaton. To compute the behaviour of a state q, we can eliminate successively all the other states of the automaton. We get an automaton with only one state q, the behaviour of which is given by the equation $\mathcal{S}(q) = o(q) + w(q)(q)\mathcal{S}(q)$, which leads to $\mathcal{S}(q) = w(q)(q)^* \times o(q)$. Since $o(q)$ and $w(q)(q)$ are given by rational expressions, this gives us a rational expression for $\mathcal{S}(q)$.

Example. Consider the automaton on the right. Removing successively q_1 and q_2 leads to:

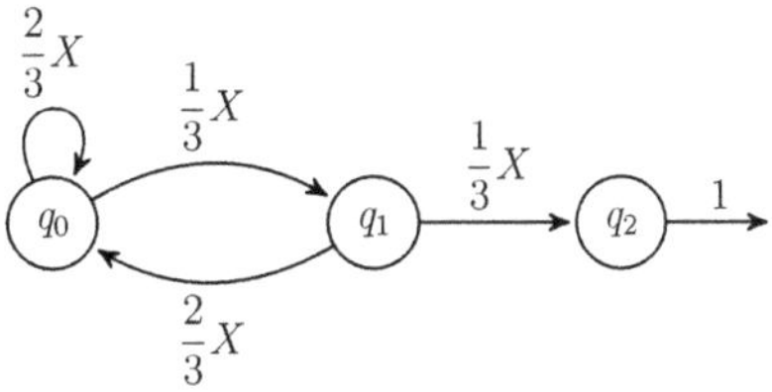

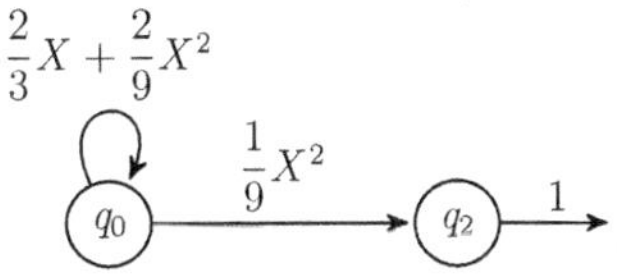

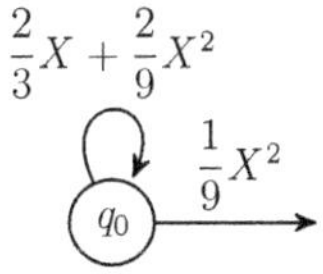

Finally, we get $\mathcal{S}(q_0) = \left(\frac{2}{3}X + \frac{2}{9}X^2\right)^* \times \frac{1}{9}X^2.$

5.2 Reduction of Well-Shaped Infinite Weighted Automata

Most of what is known about weighted automata concerns finite automata. Yet some situations can conveniently be represented by an infinite weighted automaton, in a rather natural way.

For instance, in [4,10], infinite weighted automata are extensively used to model counting problems. The idea is to give a weighted automata recognizing the generating function of the family of combinatorial objects studied, that is, the power series $\sum_{n\in\mathbb{N}} f_n X^n$, where f_n denotes the number of objects of size n. Typically, such a generating function is represented by an infinite weighted automaton in which most transitions have weight 1, and which is constructed in such a way that each object of size n correspond precisely to one accepting path of length n in the automaton.

Example. One of the main examples given in [4] is the study of Motzkin paths. A Motzkin path of length n is a lattice path of $\mathbb{Z} \times \mathbb{Z}$ going from $(0,0)$ to $(n,0)$, that never passes below the x-axis and whose permitted steps are the up diagonal step $(1,1)$, the down diagonal step $(1,-1)$ and the level step $(1,0)$.

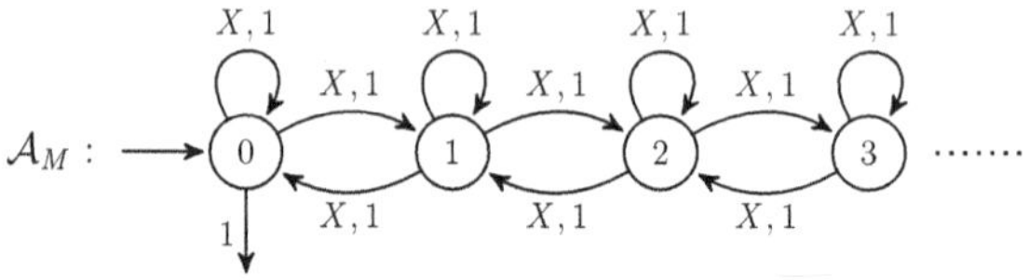

Fig. 3. Weighted automaton for Motzkin paths.

The number Motzkin paths of length n, called the n-th Motzkin number, is given by the behaviour of the automaton $\mathcal{A}_M$ (see Fig. 3); more precisely, it is equal to $\mathcal{S}(0)(n)$. Intuitively, a transition $i \to (i+1)$ corresponds to a step $(1,1)$, a transition $(i+1) \to i$ to a step $(1,-1)$, and a transition $i \to i$ to a step $(1,0)$. A Motzkin path starts and ends at level $y = 0$, which is why 0 is both the only initial and final state.

In this example as in many others, the automaton that interests us is constructed by repeating the same pattern infinitely often. Our aim is to use these regularities to contract some part of the automaton, and compute an equivalent finite HWA when it is possible.

Example. The automaton in Fig. 3 is equivalent to the automaton to the right. The idea is the following. Consider an accepting path starting in state 1, that is, a path $1 = q_0 \to q_1 \to \cdots \to q_n = 0$. Let k be the smallest index strictly greater than 0 such that $q_k = 0$. Necessarily, $q_{k-1} = 1$, and $q_0 \to \cdots \to q_{k-1}$ is a path from 1 to 1 that never passes through 0: there are as many such paths as there are paths from 0 to 0 of length $k-1$, i.e. $\mathcal{S}(0)(k-1) = X \times \mathcal{S}(0)(k)$. This is what is expressed by $1 \xrightarrow{X,\mathcal{S}(0)} 0$ in $\hat{\mathcal{A}}_M$.

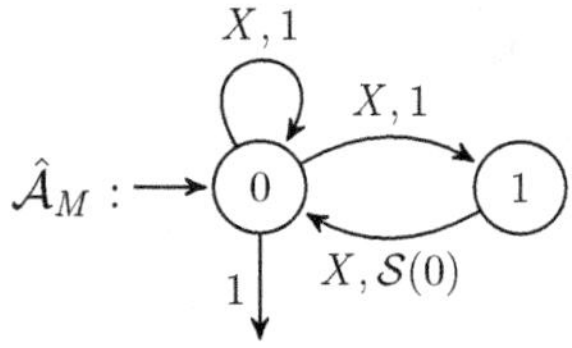

Fig. 4. Simplified automaton for Motzkin paths.

Formally, we can prove that the two automata are equivalent as follows. Denote by $\hat{\mathcal{S}}(i)$ the semantics of state i in $\hat{\mathcal{A}}_M$. We can check that the closure under linear combinations of the relation

$$\mathcal{R} = \left\{ \left(\mathcal{S}(0), \hat{\mathcal{S}}(0)\right), \left(\mathcal{S}(1), \hat{\mathcal{S}}(1)\right) \right\} \cup \left\{ \left(\mathcal{S}(i), X \times \mathcal{S}(i-1) \times \hat{\mathcal{S}}(0)\right) \mid i \geq 1 \right\}$$

is a bisimulation; hence $\mathcal{S}(0) = \hat{\mathcal{S}}(0)$ and $\mathcal{S}(1) = \hat{\mathcal{S}}(1)$.

From the automaton $\hat{\mathcal{A}}_M$, we then get the following equation for $\mathcal{S}(0)$:

$$\mathcal{S}(0) = 1 + X \times \mathcal{S}(0) + X^2 \times \mathcal{S}(0)^2.$$

A few other examples of such reductions are given in [4]. Our aim here is to identify specific classes for which such reductions are possible, and to give a few general rules applicable in these situations. The key point in the reduction we did for $\mathcal{A}_M$ was the fact that the sub-automaton consisting of the states $\{1, 2, \ldots\}$ is isomorphic to $\mathcal{A}_M$. Similarly, the rules we will give apply to automata where some sub-automaton is isomorphic to the whole automaton.

Syntactic Heavy-Weighted Automata. In the automaton $\hat{\mathcal{A}}_M$ (see Fig. 4), the stream $\mathcal{S}(0)$ appearing in the transition from 1 to 0 refers to the behaviour of the state 0 of the automaton $\mathcal{A}_M$ (see Fig. 3), and not $\hat{\mathcal{A}}_M$ itself. A priori, we can't refer to the behaviour of an automaton in the definition of its transition function, since its behaviour is itself defined using the transition function.

Yet $\hat{\mathcal{A}}_M$ is precisely constructed so as to have $\mathcal{S}(0) = \hat{\mathcal{S}}(0)$, and we would like to be able to define $\hat{\mathcal{A}}_M$ without refering to $\mathcal{A}_M$. To allow such definitions, we introduce a new kind of automata: *syntactic heavy-weighted automata*. To simplify notations, we will identify the behaviour of a state with this state itself. For instance, the syntactic heavy-weighted automaton corresponding to $\hat{\mathcal{A}}_M$ will be the one on the right.

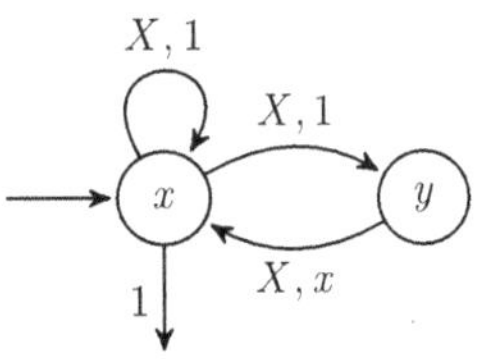

A *syntactic heavy-weighted automata* (or SHWA, for short) over a semiring S and an alphabet A consists of a pair $(Q, \langle o, t \rangle)$, where Q is the set of states, $o : Q \to S$ is the output function, and $t : Q \to (Q \to_f S\langle A \cup Q \rangle)^A$ is the transition function.

The *behaviour* $\mathcal{S}(x)$ of a state $x \in Q$ is defined as the unique solutions to the following system of behavioural differential equations: for all $x \in Q$ and $a \in A$,

$$O(x) = o(x) \qquad\qquad x_a = \sum_{y \in Q} t(x)(a)(y) \times y. \tag{1}$$

We extend $\mathcal{S} : Q \to S\langle\langle A \rangle\rangle$ to polynomials and power series as follows. We first define $\mathcal{S}$ inductively on words over $(A \cup Q)$: for all $a \in A$, $x \in Q$ and $u \in A^*$,

$$\mathcal{S}(\varepsilon) = 1 \qquad \mathcal{S}(a \cdot u) = a \times \mathcal{S}(u) \qquad \mathcal{S}(x \cdot u) = \mathcal{S}(x) \times \mathcal{S}(u).$$

For all $\sigma \in S\langle\langle A \cup Q \rangle\rangle$ and $w \in A^*$, we then define $\mathcal{S}(\sigma) = \sum_{w \in A^*} \sigma(w)\mathcal{S}(w)$. For instance, we have $\mathcal{S}(3aq + qr) = 3a\mathcal{S}(q) + \mathcal{S}(q)\mathcal{S}(r)$. The fact that $\{\mathcal{S}(x) \mid x \in Q\}$ is the solution to the system of behavioural differential equations in (1) can then be written as follows: for all $x \in Q$ and $a \in A$,

$$O(\mathcal{S}(x)) = o(x) \qquad\qquad \mathcal{S}(x)_a = \sum_{y \in Q} \mathcal{S}(t(x)(a)(y)) \times \mathcal{S}(y).$$

First Reduction Rule. We describe a method to remove a (possibly infinite) set of states from a SHWA $\mathcal{A}$, under precise assumptions. We proceed in two steps: first, we show how to disconnect a subset Q' of the states of $\mathcal{A}$ from the rest of the states (Proposition 3), and then we show that when the sub-automaton induced by Q' is isomorphic to $\mathcal{A}$, we can remove it entirely (Corollary 1).

Let $\mathcal{A} = (Q, \langle o, t\rangle)$ be a SHWA, and $Q' \subsetneq Q$. Let

$$F' = \{q \in Q' \mid \exists r \in Q \setminus Q',\ \exists a \in A,\ t(q)(a)(r) \neq 0\},$$

and

$$I' = \{q \in Q' \mid \exists r \in Q \setminus Q',\ \exists a \in A,\ t(r)(a)(q) \neq 0\}.$$

We assume that Q' satisfies the following conditions:

(1) For all $q \in Q'$, $o(q) = 0$.
(2) For all $p, q \in Q$ and $a \in A$, $t(p)(a)(q) \in S\langle A \cup (Q \setminus Q')\rangle$.
(3) There exists $o' : Q \to S$ and $f : A \to S\langle A \cup (Q \setminus Q')\rangle^{Q \setminus Q'}$ such that:
   (a) for all $q \in Q'$, $(o'(q) \neq 0 \iff q' \in F')$.
   (b) for all $q \in Q'$ and $r \in Q \setminus Q'$, $t(q)(a)(r) = o'(q) \times f(a)(r)$.

If F' contains only one state q, condition (3) becomes useless, and we simply set $o'(q) = 1$ and $f(a)(r) = t(q)(a)(r)$.

Define $\hat{\mathcal{A}} = \left(Q, \langle \hat{o}, \hat{t}\rangle\right)$ as follows:

- $\hat{o}(q) = \begin{cases} o(q) & \text{if } q \in Q \setminus Q' \\ o'(q) & \text{if } q \in Q' \end{cases}$
- $\hat{t}(p)(a)(q) = \begin{cases} t(p)(a)(q) + \sum_{r \in Q'} t(p)(a)(r) \times r & \text{if } p, q \in Q \setminus Q' \\ \quad \times \sum_{b \in A} b \times f(b)(q) & \\ t(p)(a)(q) & \text{if } p, q \in Q' \\ 0 & \text{otherwise.} \end{cases}$

We denote by $\hat{\mathcal{S}}(q)$ the behaviour of a state $q \in Q$ in automaton $\hat{\mathcal{A}}$, and by $\mathcal{S}(q)$ its behaviour in automaton $\mathcal{A}$.

Proposition 3. *Under the above assumptions $\hat{\mathcal{S}}(q) = \mathcal{S}(q)$ for all $q \in Q \setminus Q'$.*

The idea is that this construction is the opposite of the construction that we did in Sect. 4. We recognize something of the form

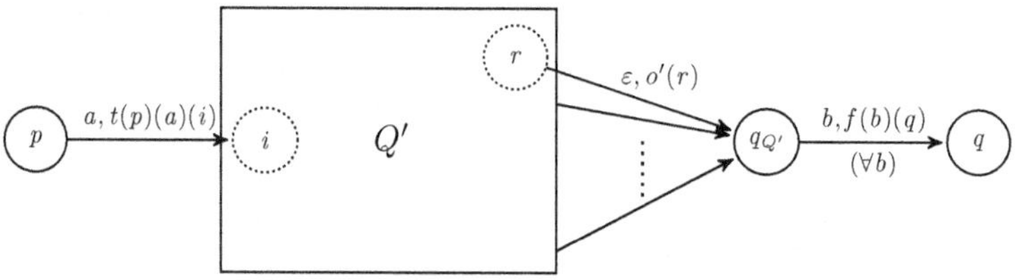

and we contract it into the one on the right which, after removal of $q_{Q'}$ (as in Sect. 5.1), leads for all $a \in A$ to a transition $p \xrightarrow{a, t(p)(a)(i) \times i \times \sum_b b f(b)(q)} q$.

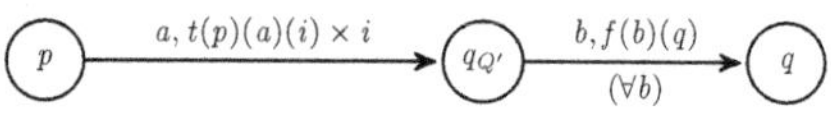

Proof (of Proposition 3). Let $C = \sum_{a \in A} \sum_{r \in Q \setminus Q'} a \times \mathcal{S}(f(a)(r)) \times \mathcal{S}(r)$. Then $\{(\mathcal{S}(q), \hat{\mathcal{S}}(q)) \mid q \in Q \setminus Q'\} \cup \{(\mathcal{S}(q), \hat{\mathcal{S}}(q) \times C) \mid q \in Q'\}$ is a bisimulation-up-to. □

Proposition 3 allows us to isolate the sub-automaton of $\mathcal{A}$ obtained by keeping only the states in Q', and setting as final those states that have an outgoing transition leaving Q'. More interestingly, when this sub-automaton is isomorphic to $\mathcal{A}$ itself, we can remove all the states in Q', as follows.

Corollary 1. *Assume that there exists a bijection $\varphi : Q' \to Q$ such that:*

- $\forall q \in Q',\ o'(q) = o(\varphi(q))$
- $\forall p, q \in Q,\ t(\varphi(p))(a)(\varphi(q)) = t(p)(a)(q)$
- $\forall q \in I',\ \varphi(q) \in Q \setminus Q'$

We define $\bar{\mathcal{A}} = (Q \setminus Q', \langle \bar{o}, \bar{t} \rangle)$ as follows: $\bar{o}$ is the restriction of o to $Q \setminus Q'$, and for all $p, q \in Q \setminus Q'$,

$$\bar{t}(p)(a)(q) = t(p)(a)(q) + \sum_{r \in Q'} t(p)(a)(r) \times \varphi(r) \times \sum_{b \in A} bf(b)(q).$$

Then $\bar{\mathcal{S}}(q) = \mathcal{S}(q)$ for all $q \in Q \setminus Q'$.

Proof. $\bar{t}$ is well-defined, because of condition (2) and the assumption that for all $q \in I', \varphi(q) \in Q \setminus Q'$. To prove the equality, we show successively that $\mathcal{R}_1 = \left\{\left(\mathcal{S}(\varphi(q)), \hat{\mathcal{S}}(q)\right) \mid q \in Q'\right\}$ and $\mathcal{R}_2 = \left\{\left(\bar{\mathcal{S}}(q), \hat{\mathcal{S}}(q)\right) \mid q \in Q \setminus Q'\right\}$ are bisimulations-up-to. Hence $\bar{\mathcal{S}}(q) = \hat{\mathcal{S}}(q) = \mathcal{S}(q)$ for all $q \in Q \setminus Q'$. □

Example. Consider again the automaton $\mathcal{A}_M$. Taking $Q' = \{q_1, q_2, \ldots\}$ and $\varphi(q_i) = q_{i-1}$, we obtain the automaton $\bar{\mathcal{A}}_M$, which can be also be obtained from $\hat{\mathcal{A}}_M$ in Fig. 4 by removing state 1.

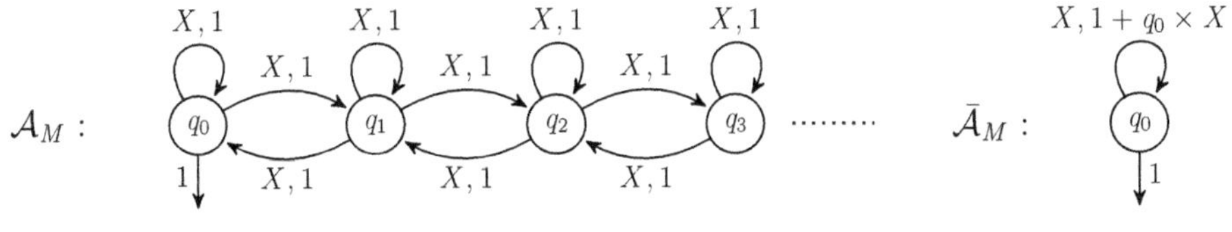

Fig. 5. Application of the first reduction rule to $\mathcal{A}_M$

Second Reduction Rule. Condition (1) in the previous reduction rule is quite restrictive, even when being interested only in specific, well-shaped automata. To gain a little more generality, we present a second reduction rule, which consists not in removing states, but in transforming some final states into non final states.

Let $\mathcal{A} = (q, \langle o, t \rangle)$ be a SHWA, and $Q' \subsetneq Q$. We define as before $I' = \{q \in Q' \mid \exists r \in Q,\ \exists a \in A,\ t(r)(a)(q) \neq 0\}$, but this time we set $F' = \{q \in Q' \mid o(q) = 0\}$.

Suppose that there exists a bijection $\psi : Q' \to Q$ such that:

(1) For all $q, r \in Q'$, $a \in A$, $t(p)(a)(q) = t(\psi(p))(a)(\psi(q))$ and $o(q) = o(\psi(q))$
(2) For all $i \in I'$, $\psi(i) \in Q \setminus Q'$.

and that

(3) For all $p, q \in Q$ and $a \in A$, $t(p)(a)(q) \in S\langle A \cup (Q \setminus Q')\rangle$

We define $\hat{\mathcal{A}} = \left(Q, \langle \hat{o}, \hat{t}\rangle\right)$ as follows: for all $p, q \in Q$ and $a \in A$,

$$\hat{o}(q) = \begin{cases} o(q) \text{ if } q \in Q \setminus Q' \\ 0 \quad\;\; \text{if } q \in Q' \end{cases}$$

$$\hat{t}(p)(a)(q) = \begin{cases} t(p)(a)(q) + t(p)(a)(\psi^{-1}(q)) \text{ if } p, q \in Q \setminus Q' \\ t(p)(a)(q) \qquad\qquad\qquad\qquad\quad \text{otherwise.} \end{cases}$$

Proposition 4. *Under the above assumptions, denote by $\mathcal{S}$ the behaviour of $\mathcal{A}$, and by $\hat{\mathcal{S}}$ the behaviour of $\hat{\mathcal{A}}$. Then $\hat{\mathcal{S}}(q) = \mathcal{S}(q)$ for all $q \in Q \setminus Q'$.*

Proof. $\left\{(\mathcal{S}(q), \hat{\mathcal{S}}(q)) \mid q \in Q \setminus Q'\right\} \cup \left\{(\mathcal{S}(q), \hat{\mathcal{S}}(q) + \mathcal{S}(\psi(q))) \mid q \in Q'\right\}$ is a bisimulation-up-to. □

Example. Consider the automaton on the right (taken from [4]), where all final states have output 1. Taking $Q' = \{q_i \mid i \geq 2\}$ and $\psi(q_i) = q_{i-2}$, we obtain:

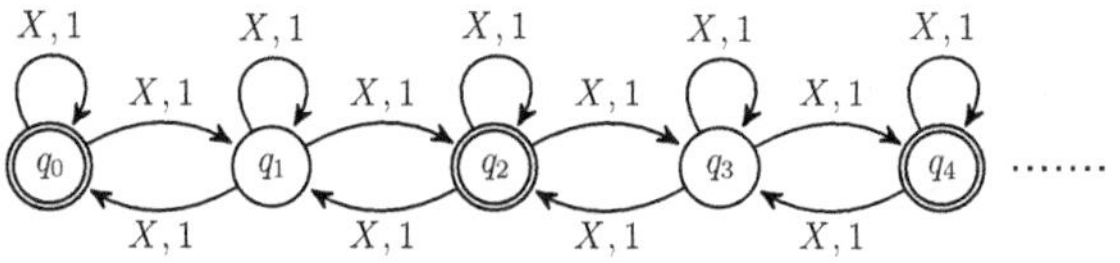

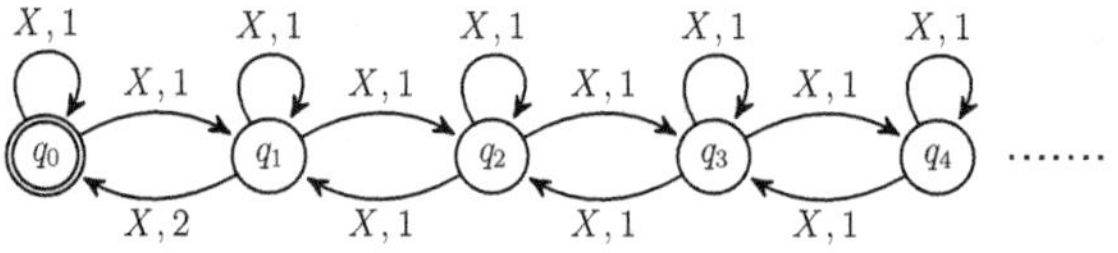

Using Proposition 3, q_0 has the same behaviour in the automaton on the right. And combining this with the reduction shown in Fig. 5, we get:

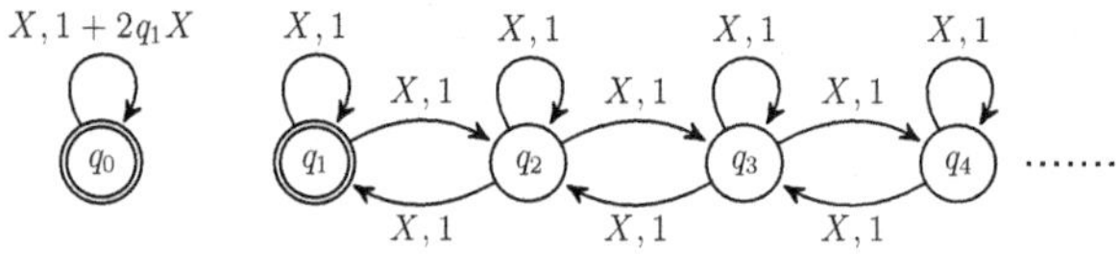

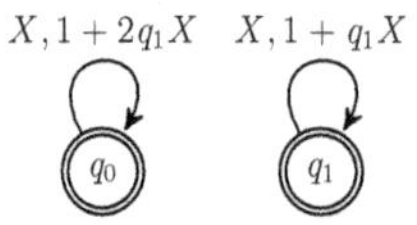

The behaviour of q_0 is then described by:

$$O(\mathcal{S}(q_0)) = 1 \qquad \mathcal{S}(q_0)' = \mathcal{S}(q_0) + 2\mathcal{S}(q_1)X\mathcal{S}(q_0)$$
$$O(\mathcal{S}(q_1) = 1 \qquad \mathcal{S}(q_1)' = \mathcal{S}(q_1) + \mathcal{S}(q_1)X\mathcal{S}(q_1).$$

Link with Algebraic Power Series. All the examples we treated in this section are *algebraic* (or *context-free*) power series. More generally, SHWAs can be seen as a representation of polynomial systems of equations. In the finite case, the solution to such a system of equations is an algebraic power series.

A *polynomial system of behavioural differential equations* over a semiring S, an alphabet A, and a set of variables $\mathcal{X}$ consists of a set of equations (one for each $x \in \mathcal{X}$) of the form

$$x_a = t \qquad\qquad O(x) = s$$

where $t \in S\langle A \cup \mathcal{X}\rangle$ and $s \in S$. A polynomial system of behavioural differential equations is *finite* if $\mathcal{X}$ is finite.

A polynomial system of behavioural differential equations always has a unique solution. A formal power series $\sigma \in S\langle\langle A\rangle\rangle$ is called *algebraic* when it is part of the solution of a *finite* polynomial system of behavioural differential equations.

This coinductive characterization of algebraic power series is equivalent to other notions of algebraic or context-free power series [7], as shown in [2].

Proposition 5. *Let $\sigma \in S\langle\langle A\rangle\rangle$. Then σ is algebraic if and only if there exists a finite* SHWA $\mathcal{A} = (Q, \langle o, t\rangle)$ *and $q_0 \in Q$ such that $\mathcal{S}(q_0) = \sigma$.*

6 Conclusion

We studied an extension of weighted automata that allows the weights of the transitions to be any power series in $S\langle\langle A\rangle\rangle$, rather than elements of S. The semantics of a heavy weighted automaton can be given by a system of behavioural differential equations linking the behaviours of the different states, or by transforming the automaton into an $S \times (-)^A$-coalgebra and applying the final $S \times (-)^A$-homomorphism. Moreover, any heavy weighted automaton can be transformed into a weighted automaton in a canonical way.

Heavy weighted automata often provide a more compact representation of a power series than weighted automata. In particular, they can be used to compute a regular expression associated with a finite weighted automata, or in some cases to give a finite representation of an infinite weighted automata. The state elimination method can be used to remove one state at a time, and in some special cases, see Sect. 5.2, allow to remove an infinite subset of states.

References

1. Bonchi, F., Bonsangue, M.M., Boreale, M., Rutten, J.J.M.M., Silva, A.: A coalgebraic perspective on linear weighted automata. Inf. Comput. **211**, 77–105 (2012)
2. Bonsangue, M.M., Rutten, J., Winter, J.: Defining context-free power series coalgebraically. In: Pattinson, D., Schröder, L. (eds.) CMCS 2012. LNCS, vol. 7399, pp. 20–39. Springer, Heidelberg (2012)
3. Brzozowski, J., Mccluskey, E.J.: Signal flow graph techniques for sequential circuit state diagrams. IEEE Trans. Electron. Comput. **12**(2), 67–76 (1963)
4. Castro, R.D., Ramírez, A., Ramírez, J.L.: Applications in enumerative combinatorics of infinite weighted automata and graphs. Sci. Annal. Comput. Sci. **24**(1), 137–171 (2014)

5. Droste, M., Kuich, W., Vogler, H. (eds.): Handbook of Weighted Automata. Monographs in Theoretical Computer Science. An EATCS Series, 1st edn. Springer, Heidelberg (2009)
6. Fortin, M., Bonsangue, M.M., Rutten, J.J.M.M.: Coalgebraic semantics of heavy-weighted automata. Technical report FM-1405, CWI - Amsterdam (2014). http://oai.cwi.nl/oai/asset/22603/22603D.pdf
7. Petre, I., Salomaa, A.: Algebraic systems and pushdown automata. In: Droste, M., Kuich, W., Vogler, H. (eds.) Handbook of Weighted Automata [5]. Monographs in Theoretical Computer Science. An EATCS Series, pp. 257–289. Springer, Heidelberg (2009)
8. Rot, J., Bonsangue, M., Rutten, J.: Coalgebraic bisimulation-up-to. In: van Emde Boas, P., Groen, F.C.A., Italiano, G.F., Nawrocki, J., Sack, H. (eds.) SOFSEM 2013. LNCS, vol. 7741, pp. 369–381. Springer, Heidelberg (2013)
9. Rutten, J.J.M.M.: Behavioural differential equations: a coinductive calculus of streams, automata, and power series. Theoret. Comput. Sci. **308**(1–3), 1–53 (2003)
10. Rutten, J.J.M.M.: Coinductive counting with weighted automata. J. Automata Lang. Comb. **8**(2), 319–352 (2003)
11. Rutten, J.J.M.M.: A coinductive calculus of streams. Math. Struct. Comput. Sci. **15**(1), 93–147 (2005)
12. Sakarovitch, J.: Elements of Automata Theory. Cambridge University Press, New York (2009)
13. Silva, A., Bonchi, F., Bonsangue, M.M., Rutten, J.J.M.M.: Generalizing determinization from automata to coalgebras. Log. Methods Comput. Sci. **9**(1) (2013)
14. Wood, D.: Theory of Computation. Harper & Row, New York (1987)

Foundations of Logic Programming in Hybridised Logics

Daniel Găină(✉)

Research Center for Software Verification, Japan Advanced Institute of Science and Technology (JAIST), Nomi, Japan
daniel@jaist.ac.jp

Abstract. The present paper sets the foundation of logic programming in hybridised logics. The basic logic programming semantic concepts such as query and solutions, and the fundamental results such as the existence of initial models and Herbrand's theorem, are developed over a very general hybrid logical system. We employ the hybridisation process proposed by Diaconescu over an arbitrary logical system captured as an institution to define the logic programming framework.

1 Introduction

Hybrid logics [1] are a brand of modal logics that allows direct reference to the possible worlds/states in a simple and very natural way through the so-called nominals. This feature has several advantages from the point of view of logic and formal specification. For example, it becomes considerably simpler to define proof systems in hybrid logics [2], and one can prove results of a generality that is not available in non-hybrid modal logic. In specifications of dynamic systems the possibility of explicit reference to specific states of the model is an essential feature.

The hybridisation of a logic is the process of developing the features of hybrid logic on top of the base logic both at the syntactic level (i.e. modalities, nominals, etc.) and semantics (i.e. possible worlds). By a hybridised institution (or hybrid institution) we mean the result of this process when logics are treated abstractly as institutions [7]. The hybridisation development in [6,13] abstracts away the details, both at the syntactic and semantic levels, that are independent of the very essence of the hybrid logic idea. One great advantage of this approach is the clarity of the theoretical developments that are not hindered by the irrelevant details of the concrete logics. Another practical benefit is the applicability of the results to a wide variety of concrete instances.

In this paper we investigate a series of model-theoretic properties of hybrid logics in an institution-independent setting such as basic set of sentences [3], substitution [4] and reachable model [10,11]. While the definition of basic set of sentences is a straightforward extension from a base institution to its hybrid counterpart, the notion of substitution needs much consideration. Establishing an appropriate concept of substitution is the most difficult part of the whole

M. Codescu et al. (Eds.): WADT 2014, LNCS 9463, pp. 69–89, 2015.
DOI: 10.1007/978-3-319-28114-8_5

enterprise of constructing an initial model of a given hybrid theory and proving a variant of Herbrand's theorem. The notion of substitution is closely related to quantification. Our abstract results are applicable to hybrid logical systems where the variables may be interpreted differently across distinct worlds, which amounts to the world-line semantics of [14]. Our paper does not cover the rigid quantification of [2] when the possible worlds share the same domain and the variables are interpreted the same in all worlds.

Initial semantics [8] is closely related to good computational properties of logical systems and it plays a crucial role for the semantics of abstract data types. For example, initiality supports the execution of specification languages through rewriting, thus integrating efficiently formal verification of software systems into modelling. The initial semantics methodology has spread much beyond its original context, that of traditional equational specification, to a variety of modern and more sophisticated logical contexts. Moreover, initial semantics plays a foundational role in logic programming. For example, in [12], initial models are known as "least Herbrand models". Our approach to initiality is layered and is intimately linked to the structure of sentences, in the style of [9]. The existence of initial models of sets of atomic sentences is assumed in abstract setting but is developed in concrete examples; then the initiality property is shown to be closed under certain sentence building operators.

The second main contribution of the paper is a variant of Herbrand's theorem for hybrid institutions, which reduces the satisfiability of a query with respect to a hybrid theory to the search of a suitable substitution. The logic programming paradigm [12], in its classical form, can be described as follows: Given a program (Σ, Γ) (that consists of a signature Σ and a set of Horn clauses Γ) and a query $(\exists Y)\rho$ (that consists of an existentially quantified conjunction of atoms) find a solution θ, i.e. values for the variables Y such that the corresponding instance $\theta(\rho)$ of ρ is satisfied by (Σ, Γ). The essence of this paradigm is however independent of any logical system of choice. The basic logic programming concepts, query, solutions, and the fundamental results, such as Herbrand's theorem, are developed over an arbitrary institution (satisfying certain hypotheses) in [4] by employing institution-independent concepts of variables, substitution, quantifiers and atomic formulas. Our work sets foundation for a uniform development of logic programming over a large variety of hybrid logics as we employ the hybridisation process over an arbitrary institution [6,13] to prove the desired results.

The institution-independent status of the present study makes the results applicable to a multitude of concrete hybrid logics including those obtained from hybridisation of non-conventional logics used in computer science.

The paper is organised as follows: in Sect. 2 we recall the definition of institution and the related notions such as substitution, reachable model and basic set of sentences. In Sect. 3 we recall the institution-indepedent process of hybridisation of a logical system and we lift the notions discussed in the previous section to the hybrid setting. Section 4 is dedicated to the development of the layered initiality result. In Sect. 5 we present an institution-independent version of Herbrand's theorem and its applications to concrete hybrid logics. Section 6 concludes the paper and discusses the future work.

2 Institutions

The concept of institution formalises the intuitive notion of logical system, and has been defined by Goguen and Burstall in the seminal paper [7].

Definition 1. *An* institution $\mathtt{I} = (\mathbb{S}ig^{\mathtt{I}}, \mathbb{S}en^{\mathtt{I}}, \mathbb{M}od^{\mathtt{I}}, \models^{\mathtt{I}})$ *consists of*

(1) a category $\mathbb{S}ig^{\mathtt{I}}$*, whose objects are called* signatures,
(2) a functor $\mathbb{S}en^{\mathtt{I}} : \mathbb{S}ig^{\mathtt{I}} \to \mathbb{S}et$*, providing for each signature* Σ *a set whose elements are called* (Σ-)sentences,
(3) a functor $\mathbb{M}od^{\mathtt{I}} : (\mathbb{S}ig^{\mathtt{I}})^{op} \to \mathbb{CAT}$*, providing for each signature* Σ *a category whose objects are called* (Σ-)models *and whose arrows are called* (Σ-)morphisms,
(4) a relation $\models^{\mathtt{I}}_{\Sigma} \subseteq |\mathbb{M}od^{\mathtt{I}}(\Sigma)| \times \mathbb{S}en^{\mathtt{I}}(\Sigma)$ *for each signature* $\Sigma \in |\mathbb{S}ig^{\mathtt{I}}|$*, called* ($\Sigma$-)satisfaction, *such that for each morphism* $\varphi : \Sigma \to \Sigma'$ *in* $\mathbb{S}ig^{\mathtt{I}}$*, the following* satisfaction condition *holds:*

$$M' \models^{\mathtt{I}}_{\Sigma'} \mathbb{S}en^{\mathtt{I}}(\varphi)(e) \text{ iff } \mathbb{M}od^{\mathtt{I}}(\varphi)(M') \models^{\mathtt{I}}_{\Sigma} e$$

for all $M' \in |\mathbb{M}od^{\mathtt{I}}(\Sigma')|$ *and* $e \in \mathbb{S}en^{\mathtt{I}}(\Sigma)$*.*

When there is no danger of confusion, we omit the superscript from the notations of the institution components; for example $\mathbb{S}ig^{\mathtt{I}}$ may be simply denoted by $\mathbb{S}ig$. We denote the *reduct* functor $\mathbb{M}od(\varphi)$ by $\_ \upharpoonright_{\varphi}$ and the sentence translation $\mathbb{S}en(\varphi)$ by $\varphi(\_)$. When $M = M' \upharpoonright_{\varphi}$ we say that M is the φ*-reduct* of M' and M' is a φ*-expansion* of M. We say that φ is *conservative* if each Σ-model has a φ-expansion. Given a signature Σ and two sets of Σ-sentences E_1 and E_2, we write $E_1 \models\!\!\!\dashv E_2$ whenever $E_1 \models E_2$ and $E_2 \models E_1$.

The literature shows myriads of logical systems from computing or mathematical logic captured as institutions (see, for example, [5]).

Example 1 (First-Order Logic (**FOL**) *[7]).* The signatures are triplets (S, F, P), where S is the set of sorts, $F = \{F_{\underline{ar}\to s})\}_{(\underline{ar},s)\in S^*\times S}$ is the ($S^*\times S$ -indexed) set of operation symbols, and $P = \{P_{\underline{ar}}\}_{\underline{ar}\in S^*}$ is the (S^*-indexed) set of relation symbols. If $\underline{ar} = \epsilon$, where ϵ denotes the empty arity, an element of $F_{\underline{ar}\to s}$ is called a *constant symbol*, or a *constant.* By a slight notational abuse, we let F and P also denote $\bigcup_{(\underline{ar},s)\in S^*\times S} F_{\underline{ar}\to s}$ and $\bigcup_{\underline{ar}\in S^*} P_{\underline{ar}}$, respectively. A signature morphism between (S, F, P) and (S', F', P') is a triplet $\varphi = (\varphi^{st}, \varphi^{op}, \varphi^{rl})$, where $\varphi^{st} : S \to S'$, $\varphi^{op} : F \to F'$, $\varphi^{rl} : P \to P'$ such that for all $(\underline{ar}, s) \in S^* \times S$ we have $\varphi^{op}(F_{\underline{ar}\to s}) \subseteq F'_{\varphi^{st}(\underline{ar})\to\varphi^{st}(s)}$, and for all $\underline{ar} \in S^*$ we have $\varphi^{rl}(P_{\underline{ar}}) \subseteq P'_{\varphi^{st}(\underline{ar})}$. When there is no danger of confusion, we may let φ denote each of φ^{st}, φ^{op}, φ^{rl}. Given a signature $\Sigma = (S, F, P)$, a Σ-model is a triplet $M = (\{s_M\}_{s\in S}, \{\sigma_M\}_{(\underline{ar},s)\in S^*\times S, \sigma\in F_{\underline{ar}\to s}}, \{\pi_M\}_{\underline{ar}\in S^*, \pi\in P_{\underline{ar}}})$ interpreting each sort s as a set s_M, each operation symbol $\sigma \in F_{\underline{ar}\to s}$ as a function $\sigma_M : \underline{ar}_M \to$

s_M (where $\underline{ar}_M$ stands for $(s_1)_M \times \ldots \times (s_n)_M$ if $\underline{ar} = s_1 \ldots s_n$), and each relation symbol $\pi \in P_{\underline{ar}}$ as a relation $\pi_M \subseteq \underline{ar}_M$. Morphisms between models are the usual Σ-morphisms, i.e., S-sorted functions that preserve the structure. The Σ-algebra of terms is denoted by T_Σ. The Σ-sentences are obtained from (a) equality atoms (e.g. $t_1 = t_2$, where $t_1, t_2 \in s_{T_\Sigma}$, $s \in S$) or (b) relational atoms (e.g. $\pi(t_1, \ldots, t_n)$, where $\pi \in P_{s_1 \ldots s_n}$, $t_i \in (s_i)_{T_\Sigma}$, $s_i \in S$ and $i \in \{1, \ldots, n\}$) by applying for a finite number of times Boolean connectives and quantification over finite sets of variables. Satisfaction is the usual first-order satisfaction and is defined using the natural interpretations of ground terms t as elements t_M in models M. The definitions of functors $\mathbb{S}en$ and $\mathbb{M}od$ on morphisms are the natural ones: for any signature morphism $\varphi : \Sigma \to \Sigma'$, $\mathbb{S}en(\varphi) : \mathbb{S}en(\Sigma) \to \mathbb{S}en(\Sigma')$ translates sentences symbol-wise, and $\mathbb{M}od(\varphi) : \mathbb{M}od(\Sigma') \to \mathbb{M}od(\Sigma)$ is the forgetful functor.

*Example 2 (***REL***).* The institution **REL** is the sub-institution of single-sorted first-order logic with signatures having only constants and relational symbols.

*Example 3 (Propositional Logic (***PL***)).* The institution **PL** is the fragment of **FOL** determined by signatures with empty sets of sort symbols.

Example 4 (Constrained Institutions). Let $\mathtt{I} = (\mathbb{S}ig^{\mathtt{I}}, \mathbb{S}en^{\mathtt{I}}, \mathbb{M}od^{\mathtt{I}}, \models^{\mathtt{I}})$ be an institution. A *constrained model functor* $\mathbb{M}od^{\mathtt{CI}} : (\mathbb{S}ig^{\mathtt{CI}})^{op} \to \mathbb{CAT}$ is a sub-functor of $\mathbb{M}od^{\mathtt{I}} : (\mathbb{S}ig^{\mathtt{I}})^{op} \to \mathbb{CAT}$, i.e. $\mathbb{S}ig^{\mathtt{CI}} \subseteq \mathbb{S}ig^{\mathtt{I}}$, for each $\Sigma \in |\mathbb{S}ig^{\mathtt{CI}}|$ we have $\mathbb{M}od^{\mathtt{CI}}(\Sigma) \subseteq \mathbb{M}od^{\mathtt{I}}(\Sigma)$, and for each $\Sigma \xrightarrow{\varphi} \Sigma \in \mathbb{S}ig^{\mathtt{CI}}$ the functor $\mathbb{M}od^{\mathtt{CI}}(\varphi) : \mathbb{M}od^{\mathtt{CI}}(\Sigma') \to \mathbb{M}od^{\mathtt{CI}}(\Sigma)$ is defined by $\mathbb{M}od^{\mathtt{CI}}(\varphi)(h) = \mathbb{M}od^{\mathtt{I}}(\varphi)(h)$ for all $h \in \mathbb{M}od^{\mathtt{CI}}(\Sigma')$. We say that $\mathtt{CI} = (\mathbb{S}ig^{\mathtt{CI}}, \mathbb{S}en^{\mathtt{CI}}, \mathbb{M}od^{\mathtt{CI}}, \models^{\mathtt{CI}})$ is *a constrained institution*, where (a) $\mathbb{S}en^{\mathtt{CI}} : \mathbb{S}ig^{\mathtt{CI}} \to \mathbb{S}et$ is the restriction of $\mathbb{S}en^{\mathtt{I}} : \mathbb{S}ig^{\mathtt{I}} \to \mathbb{S}et$ to $\mathbb{S}ig^{\mathtt{CI}}$, and (b) $\models^{\mathtt{CI}}_\Sigma \subseteq |\mathbb{M}od^{\mathtt{CI}}(\Sigma)| \times \mathbb{S}en^{\mathtt{I}}(\Sigma)$ is the restriction of $\models^{\mathtt{I}}_\Sigma \subseteq |\mathbb{M}od^{\mathtt{I}}(\Sigma)| \times \mathbb{S}en^{\mathtt{I}}(\Sigma)$ to $|\mathbb{M}od^{\mathtt{CI}}(\Sigma)|$ for all $\Sigma \in |\mathbb{S}ig^{\mathtt{CI}}|$.

2.1 Quantification Subcategory

Let $\mathtt{I} = (\mathbb{S}ig, \mathbb{S}en, \mathbb{M}od, \models)$ be an institution. A *broad subcategory*[1] $\mathcal{Q} \subseteq \mathbb{S}ig$ is called *quantification subcategory* [6] when for each $\Sigma \xrightarrow{\chi} \Sigma' \in \mathcal{Q}$ and $\Sigma \xrightarrow{\varphi} \Sigma_1 \in \mathbb{S}ig$ there is a designated pushout

$$\begin{array}{ccc} \Sigma' & \xrightarrow{\varphi[\chi]} & \Sigma'_1 \\ {\scriptstyle\chi}\uparrow & & \uparrow{\scriptstyle\chi(\varphi)} \\ \Sigma & \xrightarrow[\varphi]{} & \Sigma_1 \end{array}$$

with $\chi(\varphi) \in \mathcal{Q}$ which is a *weak amalgamation square*[2] and such that the horizontal composition of such designated pushouts is again a designated pushout, i.e. $\chi(1_\Sigma) = \chi$,

[1] A category $\mathcal{C}$ is a broad subcategory of $\mathcal{C}'$ if $\mathcal{C}$ is a subcategory of $\mathcal{C}'$ and $\mathcal{C}$ contains all objects of $\mathcal{C}'$, i.e. $|\mathcal{C}| = |\mathcal{C}'|$.

[2] For all $M' \in |\mathbb{M}od(\Sigma')|$ and $M_1 \in |Mod(\Sigma_1)|$ such that $M'\upharpoonright_\chi = M_1 \upharpoonright_\varphi$ there exists $M'_1 \in |\mathbb{M}od(\Sigma'_1)|$ such that $M'_1 \upharpoonright_{\varphi[\chi]} = M'$ and $M'_1 \upharpoonright_{\chi(\varphi)} = M_1$.

$1_\Sigma[\chi] = 1_{\Sigma'}$, and for the following pushouts 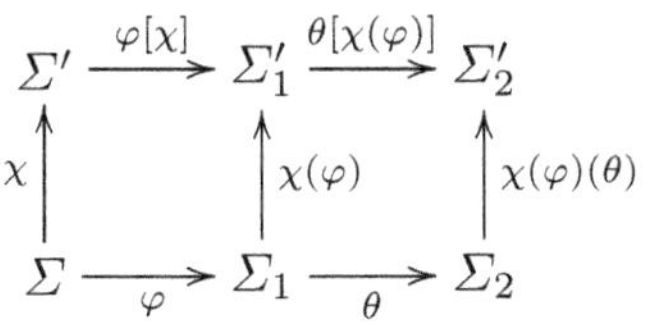 we have $\varphi[\chi];\theta[\chi(\varphi)] = (\varphi;\theta)[\chi]$ and $\chi(\varphi)(\theta) = \chi(\varphi;\theta)$.

A variable for a **FOL** signature $\Sigma = (S, F, P)$ is a triple (x, s, Σ), where x is the name of the variable and $s \in S$ is the sort of the variable. Let $\chi : \Sigma \hookrightarrow \Sigma[X]$ be a signature extension with variables from X, where $X = \{X_s\}_{s\in S}$ is a S-sorted set of variables, $\Sigma[X] = (S, F \cup X, P)$ and for all $(\underline{ar}, s) \in S^* \times S$ we have $(F \cup X)_{\underline{ar}\to s} = \begin{cases} F_{\underline{ar}\to s} & \text{if } \underline{ar} \in S^+, \\ F_{\underline{ar}\to s} \cup X_s & \text{if } \underline{ar} = \epsilon. \end{cases}$ The quantification subcategory $\mathcal{Q}^{\mathbf{FOL}}$ for **FOL** consists of signature extensions with a finite set of variables. Given a signature morphism $\varphi : \Sigma \to \Sigma_1$, where $\Sigma_1 = (S_1, F_1, P_1)$, then

- $\chi(\varphi) : \Sigma_1 \hookrightarrow \Sigma_1[X^\varphi]$, where $X^\varphi = \{(x, \varphi(s), \Sigma_1) \mid (x, s, \Sigma) \in X\}$,
- $\varphi[\chi]$ is the canonical extension of φ that maps each (x, s, Σ) to $(x, \varphi(s), \Sigma_1)$.

It is straightforward to check that $\mathcal{Q}^{\mathbf{FOL}}$ defined above is a quantification subcategory.

2.2 Substitutions

We recall the notion of substitution in institutions.

Definition 2 [4]. *Let* $\mathbb{I} = (\mathbb{S}ig, \mathbb{S}en, \mathbb{M}od, \models)$ *be an institution and* $\Sigma \in |\mathbb{S}ig|$. *For any signature morphisms* $\chi_1 : \Sigma \to \Sigma_1$ *and* $\chi_2 : \Sigma \to \Sigma_2$, *a* Σ-substitution $\theta : \chi_1 \to \chi_2$ *consists of a pair* $(\mathbb{S}en(\theta), \mathbb{M}od(\theta))$, *where*

- $\mathbb{S}en(\theta) : \mathbb{S}en(\Sigma_1) \to \mathbb{S}en(\Sigma_2)$ *is a function and*
- $\mathbb{M}od(\theta) : \mathbb{M}od(\Sigma_2) \to \mathbb{M}od(\Sigma_1)$ *is a functor.*

such that both of them preserve Σ, *i.e. the following diagrams commute:*

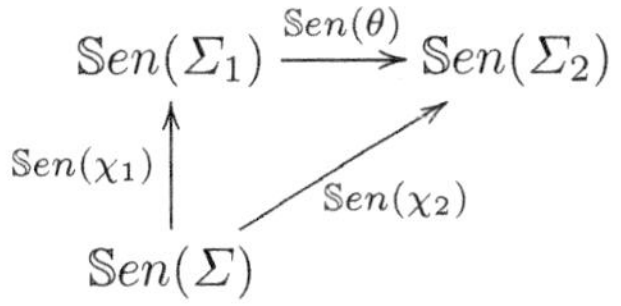

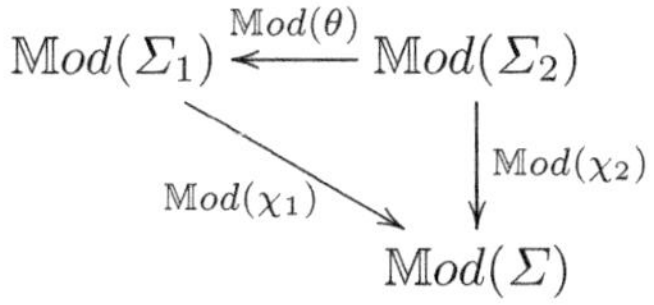

and such that the following satisfaction condition *holds:*

$$\mathbb{M}od(\theta)(M_2) \models \rho_1 \text{ iff } M_2 \models \mathbb{S}en(\theta)(\rho_1)$$

for each Σ_2*-model* M_2 *and each* Σ_1*-sentence* ρ_1.

Note that a substitution $\theta : \chi_1 \to \chi_2$ is uniquely identified by its domain χ_1, codomain χ_2 and the pair $(\mathbb{S}en(\theta), \mathbb{M}od(\theta))$. We sometimes let $\_\restriction_\theta$ denote the functor $\mathbb{M}od(\theta)$, and let θ denote the sentence translation $\mathbb{S}en(\theta)$.

*Example 5 (**FOL** substitutions [4]).* Consider two signature extensions with constants $\chi_1 : \Sigma \hookrightarrow \Sigma[C_1]$ and $\chi_2 : \Sigma \hookrightarrow \Sigma[C_2]$, where $\Sigma = (S, F, P) \in |\mathbb{S}ig^{\mathbf{FOL}}|$, C_i is a set of constant symbols different from the symbols in Σ. A function $\theta : C_1 \to T_\Sigma(C_2)$ represents a substitution between χ_1 and χ_2. On the syntactic side, θ can be canonically extended to a function $\mathbb{S}en(\theta) : \mathbb{S}en(\Sigma[C_1]) \to \mathbb{S}en(\Sigma[C_2])$ as follows:

- $\mathbb{S}en(\theta)(t_1 = t_2)$ is defined as $\theta^{term}(t) = \theta^{term}(t')$ for each $\Sigma[C_1]$-equation $t_1 = t_2$, where $\theta^{term} : T_\Sigma(C_1) \to T_\Sigma(C_2)$ is the unique extension of θ to a Σ-morphism.
- $\mathbb{S}en(\theta)(\pi(t_1, \ldots, t_n))$ is defined as $\pi(\theta^{term}(t_1), \ldots, \theta^{term}(t_n))$ for each $\Sigma[C_1]$-relational atom $\pi(t_1, \ldots, t_n)$.
- $\mathbb{S}en(\theta)(\bigwedge E)$ is defined as $\bigwedge \mathbb{S}en(\theta)(E)$ for each conjuction $\bigwedge E$ of $\Sigma[C_1]$-sentences, and similarly for the case of any other Boolean connectives.
- $\mathbb{S}en(\theta)((\forall X)\rho)$ is defined as $(\forall X^\theta)\mathbb{S}en(\theta')(\rho)$ for each $\Sigma[C_1]$-sentence $(\forall X)\rho$, where $X^\theta = \{(x, s, \Sigma[C_2]) \mid (x, s, \Sigma[C_1]) \in X\}$ and the substitution $\theta' : C_1 \cup X \to T_\Sigma(C_2 \cup X^\theta)$ extends θ by mapping each variable $(x, s, \Sigma[C_1]) \in X$ to $(x, s, \Sigma[C_2]) \in X^\theta$.

On the semantics side, θ determines a functor $\mathbb{M}od(\theta)$ between $\mathbb{M}od(\Sigma[C_2])$ and $\mathbb{M}od(\Sigma[C_1])$ such that for all $\Sigma[C_2]$-models M we have

- $\mathbb{M}od(\theta)(M)_x = M_x$, for each sort $x \in S$, or operation symbol $x \in F$, or relation symbol $x \in P$, and
- $\mathbb{M}od(\theta)(M)_x = M_{\theta(x)}$ for each $x \in C_1$.

Category of Substitutions. Let $\mathtt{I} = (\mathbb{S}ig, \mathbb{S}en, \mathbb{M}od, \models)$ be an institution and $\Sigma \in |\mathbb{S}ig|$ a signature. Σ-substitutions form a category $\mathbb{S}ubst^{\mathtt{I}}(\Sigma)$, where the objects are signature morphisms $\Sigma \xrightarrow{\chi} \Sigma' \in |\Sigma/\mathbb{S}ig|$, and the arrows are substitutions $\theta : \chi_1 \to \chi_2$ as described in Definition 2. For any substitutions $\theta : \chi_1 \to \chi_2$ and $\theta' : \chi_2 \to \chi_3$ the composition $\theta; \theta'$ consists of the pair $(\mathbb{S}en(\theta; \theta'), \mathbb{M}od(\theta; \theta'))$, where $\mathbb{S}en(\theta; \theta') = \mathbb{S}en(\theta); \mathbb{S}en(\theta')$ and $\mathbb{M}od(\theta; \theta') = \mathbb{M}od(\theta'); \mathbb{M}od(\theta)$.

Given a signature morphism $\varphi : \Sigma_0 \to \Sigma$ there exists a reduct functor $\mathbb{S}ubst^{\mathtt{I}}(\varphi) : \mathbb{S}ubst^{\mathtt{I}}(\Sigma) \to \mathbb{S}ubst^{\mathtt{I}}(\Sigma_0)$ that maps any Σ-substitution $\theta : \chi_1 \to \chi_2$ to the Σ_0-substitution $\mathbb{S}ubst(\varphi)(\theta) : \varphi; \chi_1 \to \varphi; \chi_2$ such that $\mathbb{S}en(\mathbb{S}ubst^{\mathtt{I}}(\varphi)(\theta)) = \mathbb{S}en(\theta)$ and $\mathbb{M}od(\mathbb{S}ubst^{\mathtt{I}}(\varphi)(\theta)) = \mathbb{M}od(\theta)$. It follows that $\mathbb{S}ubst^{\mathtt{I}} : \mathbb{S}ig^{op} \to \mathbb{CAT}$ is a functor. In applications not all substitutions are of interest, and it is often assumed a substitution sub-functor $\mathbb{S}ub^{\mathtt{I}} : \mathcal{D}^{op} \to \mathbb{CAT}$ of $\mathbb{S}ubst^{\mathtt{I}} : \mathbb{S}ig^{op} \to \mathbb{CAT}$ to work with, where $\mathcal{D} \subseteq \mathbb{S}ig$ is a subcategory of signature morphisms. When there is no danger of confusion we may drop the superscript $\mathtt{I}$ from the notations; for example $\mathbb{S}ub^{\mathtt{I}}$ may be simply denoted by $\mathbb{S}ub$.

*Example 6 (**FOL** substitution functor).* Given a signature $\Sigma \in |\mathbb{S}ig^{\mathbf{FOL}}|$, only Σ-substitutions represented by functions $\theta : C_1 \to T_\Sigma(C_2)$ are relevant for the present study, where C_1 and C_2 are finite sets of new constants for Σ. Let $\mathbb{S}ub^{\mathbf{FOL}} : (\mathcal{D}^{\mathbf{FOL}})^{op} \to \mathbb{CAT}$ denote the substitution functor which maps each signature Σ to the subcategory of Σ-substitutions represented by functions of the form $\theta : C_1 \to T_\Sigma(C_2)$ as above.

Example 7. (**PL** substitution functor) Let $\mathcal{D}^{\mathbf{PL}}$ be the subcategory of **PL** signature morphisms consisting of identities, and $\mathbb{S}ub^{\mathbf{PL}} : (\mathcal{D}^{\mathbf{PL}})^{op} \to \mathbb{CAT}$ the trivial substitution functor consisting also of identities.

2.3 Reachable Models

This subsection is devoted to the institution-independent characterisation of the models that consist of interpretations of terms.

Definition 3. *Let* $\mathrm{I} = (\mathbb{S}ig, \mathbb{S}en, \mathbb{M}od, \models)$ *be an institution,* $\mathcal{D} \subseteq \mathbb{S}ig$ *a broad subcategory of signature morphisms, and* $\mathbb{S}ub : \mathcal{D}^{op} \to \mathbb{CAT}$ *a substitution functor. A model* $M \in |\mathbb{M}od(\Sigma)|$*, where* $\Sigma \in |\mathbb{S}ig|$*, is* $\mathbb{S}ub$-reachable *if for every signature morphism* $\Sigma \xrightarrow{\chi} \Sigma' \in \mathcal{D}$ *and each* χ*-expansion* M' *of* M *there exists a substitution* $\theta : \chi \to 1_\Sigma \in \mathbb{S}ub(\Sigma)$ *such that* $M\restriction_\theta = M'$.

This notion of reachable model is the parametrisation of the one in [10] with substitutions.

Proposition 1. *In* **FOL***, a model is* $\mathbb{S}ub^{\mathbf{FOL}}$*-reachable iff its elements consist of interpretations of terms.*

The proof of Proposition 1 is a slight generalisation of the one in [10]. Note that in **PL**, all models are $\mathbb{S}ub^{\mathbf{PL}}$-reachable.

2.4 Basic Sentences

A set of sentences $B \subseteq Sen(\Sigma)$ is *basic* [3] if there exists a Σ-model M^B such that, for all Σ-models M, $M \models B$ iff there exists a morphism $M^B \to M$. We say that M^B is a *basic model* of B. If in addition the morphism $M^B \to M$ is unique then the set B is called *epi basic*; in this case, M^B is the initial model of B.

Lemma 1. *Any set of atoms in* **FOL** *is epi basic and the corresponding basic models consist of interpretations of terms, i.e. are* $\mathbb{S}ub^{\mathbf{FOL}}$*-reachable.*

Proof. Let B be a set of atomic (S, F, P)-sentences in **FOL**. The basic model M^B is the initial model of B and it is constructed as follows: on the quotient $T_{(S,F)}/_{\equiv_B}$ of the term model $T_{(S,F)}$ by the congruence generated by the equational atoms of B, we interpret each relation symbol $\pi \in P$ by $\pi_{M^B} = \{(\widehat{t_1}, \ldots, \widehat{t_n}) \mid \pi(t_1, \ldots, t_n) \in B\}$, where $\widehat{t}$ is the congruence class of t for all terms $t \in T_{(S,F)}$. □

The proof of Lemma 1 is well known, and it can be found, for example, in [3] or [5], but since it constitutes the foundation of the initiality property, we include it for the convenience of the reader. Since **PL** is obtained from **FOL** by restricting the category of signatures, every set of **PL** atoms is epi basic.

3 Hybrid Institutions

We recall the institution-independent process of hybridisation that has been introduced in [6,13]. Consider an institution $\mathtt{I} = (\mathbb{S}ig, \mathbb{S}en, \mathbb{M}od, \models)$ with a quantification subcategory $\mathcal{Q} \subseteq \mathbb{S}ig$.

The Category of HI Signatures. The category of hybrid signatures of $\mathbb{S}ig$ is defined as the following cartesian product of categories: $\mathbb{S}ig^{\mathtt{HI}} = \mathbb{S}ig^{\mathtt{I}} \times \mathbb{S}ig^{\mathbf{REL}}$. The **REL** signatures are denoted by (Nom, Λ), where Nom is a set of constants called nominals and Λ is a set of relational symbols called modalities; Λ_n stands for the set of modalities of arity n. Hybrid signatures morphisms $\varphi = (\varphi_{\mathbb{S}ig}, \varphi_{\mathrm{Nom}}, \varphi_{\mathbb{R}el}) : (\Sigma, Nom, \Lambda) \to (\Sigma', Nom, \Lambda')$ are triples such that $\varphi_{\mathbb{S}ig} : \Sigma \to \Sigma' \in \mathbb{S}ig^{\mathtt{I}}$ and $(\varphi_{\mathrm{Nom}}, \varphi_{\mathbb{R}el}) : (Nom, \Lambda) \to (Nom', \Lambda') \in \mathbb{S}ig^{\mathbf{REL}}$. When there is no danger of confusion we may drop the subscripts from notations and denote $\varphi_{\mathbb{S}ig}$, φ_{Nom} and $\varphi_{\mathbb{R}el}$ simply by φ.

HI Sentences. Let us denote by $\mathcal{Q}^{\mathtt{HI}}$ the subcategory $\mathcal{Q}^{\mathtt{HI}} \subseteq \mathbb{S}ig^{\mathtt{HI}}$ which consists of signature morphisms of the form $\chi : (\Sigma, Nom, \Lambda) \to (\Sigma', Nom, \Lambda)$ such that $\chi_{\mathbb{S}ig} \in \mathcal{Q}$, $\chi_{\mathrm{Nom}} = 1_{Nom}$ and $\chi_{\mathbb{R}el} = 1_{\Lambda}$.

Theorem 1 [6,13]. *If $\mathcal{Q}$ is a quantification subcategory for* I *then $\mathcal{Q}^{\mathtt{HI}}$ is a quantification subcategory for* HI*.*

The satisfaction condition for hybridised institutions relies upon Theorem 1. A nominal variable for a hybrid signature $\Delta = (\Sigma, Nom, \Lambda)$ is a pair of the form (x, Δ), where x is the name of the variable and Δ is the qualification of the variable. Given a hybrid signature $\Delta = (\Sigma, Nom, \Lambda)$, the set of sentences $\mathbb{S}en^{\mathtt{HI}}(\Delta)$ is the least set such that

- $Nom \subseteq \mathbb{S}en^{\mathtt{HI}}(\Delta)$,
- $\lambda(k_1, \ldots, k_n) \in \mathbb{S}en^{\mathtt{HI}}(\Delta)$ for any $\lambda \in \Lambda_{n+1}$, $k_i \in Nom$, $i \in \{1, \ldots, n\}$;
- $\mathbb{S}en^{\mathtt{I}} \subseteq \mathbb{S}en(\Delta)$;
- $\rho_1 \star \rho_2 \in \mathbb{S}en^{\mathtt{HI}}(\Delta)$ for any $\rho_1, \rho_2 \in \mathbb{S}en^{\mathtt{HI}}(\Delta)$ and $\star \in \{\wedge, \Rightarrow\}$;
- $\neg\rho \in \mathbb{S}en^{\mathtt{HI}}(\Delta)$ for any $\rho \in \mathbb{S}en^{\mathtt{HI}}(\Delta)$;
- $@_k\rho \in \mathbb{S}en^{\mathtt{HI}}(\Delta)$ for any $\rho \in \mathbb{S}en^{\mathtt{HI}}(\Delta)$ and $k \in Nom$;
- $[\lambda](\rho_1, \ldots, \rho_n)$ for any $\lambda \in \Lambda_{n+1}$, $\rho_i \in \mathbb{S}en^{\mathtt{HI}}(\Delta)$ and $i \in \{1, \ldots, n\}$;
- $(\forall\chi)\rho' \in \mathbb{S}en^{\mathtt{HI}}(\Delta)$ for any $\chi : (\Sigma, Nom, \Lambda) \to (\Sigma', Nom, \Lambda) \in \mathcal{Q}^{\mathtt{HI}}$ and $\rho' \in \mathbb{S}en^{\mathtt{HI}}(\Sigma', Nom, \Lambda)$;
- $(\forall J)\rho$ for any set J of nominal variables for Δ and $\rho \in \mathbb{S}en^{\mathtt{HI}}(\Sigma, Nom \cup J, \Lambda)$;
- $(\downarrow j)\rho$ for any nominal variable j for Δ and $\rho \in \mathbb{S}en^{\mathtt{HI}}(\Sigma, Nom \cup \{j\}, \Lambda)$.

Translation of HI Sentences. Let $\varphi : (\Sigma, Nom, \Lambda) \to (\Sigma', Nom', \Lambda')$ be a morphism of HI signatures. The translation $\mathbb{S}en^{\mathtt{HI}}(\varphi)$ is defined as follows:

- $\mathbb{S}en^{\mathtt{HI}}(\varphi)(k) = \varphi_{\mathrm{Nom}}(k)$;
- $\mathbb{S}en^{\mathtt{HI}}(\varphi)(\lambda(k_1, \ldots, k_n)) = \varphi_{\mathbb{R}el}(\lambda)(\varphi_{\mathrm{Nom}}(k_1), \ldots, \varphi_{\mathrm{Nom}}(k_n))$ for $\lambda \in \Lambda_{n+1}$, $k_i \in Nom$, $i \in \{1, \ldots, n\}$;
- $\mathbb{S}en^{\mathtt{HI}}(\varphi)(\rho) = \mathbb{S}en^{\mathtt{I}}(\varphi_{\mathbb{S}ig})(\rho)$ for any $\rho \in \mathbb{S}en^{\mathtt{I}}(\Sigma)$;
- $\mathbb{S}en^{\mathtt{HI}}(\rho_1 \star \rho_2) = \mathbb{S}en^{\mathtt{HI}}(\rho_1) \star \mathbb{S}en^{\mathtt{HI}}(\rho_2)$, where $\star \in \{\wedge, \Rightarrow\}$;

- $\mathbb{S}en^{\mathtt{HI}}(\neg\rho) = \neg\mathbb{S}en^{\mathtt{HI}}(\rho)$;
- $\mathbb{S}en^{\mathtt{HI}}(@_k\rho) = @_{\varphi_{\mathbb{N}om}(k)}\mathbb{S}en^{\mathtt{HI}}(\rho)$;
- $\mathbb{S}en^{\mathtt{HI}}([\lambda](\rho_1,\ldots,\rho_n)) = [\varphi_{\mathbb{R}el}(\lambda)](\mathbb{S}en^{\mathtt{HI}}(\rho_1),\ldots,\mathbb{S}en^{\mathtt{HI}}(\rho_n))$;
- $\mathbb{S}en^{\mathtt{HI}}((\forall\chi)\rho') = (\forall\chi(\varphi))\mathbb{S}en^{\mathtt{HI}}(\varphi[\chi])(\rho')$, where the signature morphisms $\chi : (\Sigma, Nom, \Lambda) \to (\Sigma', Nom, \Lambda)$ is in $\mathcal{Q}^{\mathtt{HI}}$, $\chi(\varphi) = (\chi_{\mathbb{S}ig}(\varphi_{\mathbb{S}ig}), 1_{Nom'}, 1_{\Lambda'})$ and $\varphi[\chi] = (\varphi_{\mathbb{S}ig}[\chi_{\mathbb{S}ig}], \varphi_{\mathbb{N}om}, \varphi_{\mathbb{R}el})$;
- $\mathbb{S}en^{\mathtt{HI}}((\forall J)\rho) = (\forall J^{\varphi})\mathbb{S}en^{\mathtt{HI}}(\varphi[J])(\rho)$, where $J^{\varphi} = \{(x,(\Sigma', Nom', \Lambda')) \mid (x,(\Sigma, Nom, \Lambda)) \in J\}$ and $\varphi[J] : (\Sigma, Nom \cup J, \Lambda) \to (\Sigma, Nom' \cup J^{\varphi}, \Lambda')$ is canonical extension of φ that maps each variable $(x,(\Sigma, Nom, \Lambda)) \in J$ to $(x,(\Sigma', Nom', \Lambda'))$;
- $\mathbb{S}en^{\mathtt{HI}}((\downarrow j)\rho) = (\downarrow j^{\varphi})\mathbb{S}en^{\mathtt{HI}}(\varphi[j])(\rho)$, where $j^{\varphi} = (x,(\Sigma', Nom', \Lambda'))$ and $\varphi^j : (\Sigma, Nom \cup \{j\}, \Lambda) \to (\Sigma', Nom' \cup \{j^{\varphi}\}, \Lambda')$ is the canonical extension of φ mapping each j to j^{φ}.

HI Models. The (Σ, Nom, Λ)-models are paris $(\mathcal{M}, R)$ where

- R is a (Nom, Λ)-model in **REL**. The carrier set $|R|$ forms the set of states of the model $(\mathcal{M}, R)$. The relations $\{\lambda_R \mid \lambda \in \Lambda_n, n \in \mathbb{N}\}$ represent the interpretation of the modalities Λ.
- $\mathcal{M}$ is a function $|R| \to \mathbb{M}od^{\mathtt{I}}(\Sigma)$. For each $s \in |R|$, we denote $\mathcal{M}(s)$ simply by $\mathcal{M}_s$.

A (Σ, Nom, Λ)-homomorphism $h : (\mathcal{M}, R) \to (\mathcal{M}', R')$ consists of

- a (Nom, Λ)-homomorphism in **REL**, $h^{st} : R \to R'$, and
- a natural transformation $h^{mod} : \mathcal{M} \Rightarrow \mathcal{M}' \circ h^{st}$.[3]

When there is no danger of confusion we may drop the superscripts *st* and *mod* from the notations h^{st} and h^{mod}, respectively. The composition of HI homomorphisms is defined canonically as $h_1; h_2 = ((h_1^{st}; h_2^{st}), h_1^{mod}; (h_2^{mod} \circ h_1^{st}))$.

Reducts of HI Models. Let $\Delta = (\Sigma, Nom, \Lambda)$ and $\Delta' = (\Sigma', Nom', \Lambda')$ be two HI signatures, $\Delta \xrightarrow{\varphi} \Delta'$ a HI signature morphism, and $(\mathcal{M}', R')$ a Δ'-model. The reduct $(\mathcal{M}, R) = \mathbb{M}od^{\mathtt{HI}}(\varphi)(\mathcal{M}', R')$ of $(\mathcal{M}', R')$ along φ denoted by $(\mathcal{M}', R')\upharpoonright_{\varphi}$, is the Δ-model such that $|R| = |R'|$, $k_R = \varphi_{\mathbb{N}om}(k)_{R'}$ for all $k \in Nom$, $\lambda_R = \varphi_{\mathbb{R}el}(\lambda)_{R'}$ for all $\lambda \in \Lambda$, and $\mathcal{M}_s = \mathbb{M}od^{\mathtt{I}}(\varphi_{\mathbb{S}ig})(\mathcal{M}'_s)$ for all $s \in |R|$.

Satisfaction Relation. For any signaturel $\Delta = (\Sigma, Nom, \Lambda)$, model $(\mathcal{M}, R) \in |\mathbb{M}od^{\mathtt{HI}}(\Delta)|$ and state $s \in |R|$ we define:

- $(\mathcal{M}, R) \models^s k$ iff $k_R = s$, for any $k \in Nom$;
- $(\mathcal{M}, R) \models^s \lambda(k_1,\ldots,k_n)$ iff $(s,(k_1)_R,\ldots,(k_n)_R) \in \lambda_R$, for any $\lambda \in \Lambda_{n+1}$, $k_i \in Nom$, $i \in \{1,\ldots,n\}$;
- $(\mathcal{M}, R) \models^s \rho$ iff $\mathcal{M}_s \models^{\mathtt{I}} \rho$ for any $\rho \in \mathbb{S}en^{\mathtt{I}}(\Sigma)$;
- $(\mathcal{M}, R) \models^s \rho_1 \wedge \rho_2$ iff $(\mathcal{M}, R) \models^s \rho_1$ and $(\mathcal{M}, R) \models^s \rho_2$;
- $(\mathcal{M}, R) \models^s \rho_1 \Rightarrow \rho_2$ iff $(\mathcal{M}, R) \models^s \rho_1$ implies $(\mathcal{M}, R) \models^s \rho_2$;

[3] h^{mod} is a $|R|$-indexed family of Σ-homomorphisms $h^{mod} = \{h_s^{mod} : \mathcal{M}_s \to \mathcal{M}'_{h^{st}(s)}\}_{s\in|R|}$.

- $(\mathcal{M}, R) \models^s \neg\rho$ iff $(\mathcal{M}, R) \not\models^s \rho$;
- $(\mathcal{M}, R) \models^s @_k\rho$ iff $(\mathcal{M}, R) \models^{k_R} \rho$;
- $(\mathcal{M}, R) \models^s [\lambda](\rho_1, \ldots, \rho_n)$ iff for every $(s, s_1, \ldots, s_n) \in \lambda_R$, $(\mathcal{M}, R) \models^{s_i} \rho_i$ for some $i \in \{1, \ldots, n\}$;
- $(\mathcal{M}, R) \models^s (\forall\chi)\rho$ iff for every expansion $(\mathcal{M}', R)$ along $\chi : (\Sigma, Nom, \Lambda) \to (\Sigma', Nom, \Lambda)$ we have $(\mathcal{M}', R) \models^s \rho$;
- $(\mathcal{M}, R) \models^s (\forall J)\rho$ iff for every expansion $(\mathcal{M}, R')$ along $\iota_J : (\Sigma, Nom, \Lambda) \hookrightarrow (\Sigma, Nom \cup J, \Lambda)$ we have $(\mathcal{M}, R') \models^s \rho$;
- $(\mathcal{M}, R) \models^s (\downarrow j)\rho$ iff $(\mathcal{M}, R') \models^s \rho$, where $(M\ R')$ is the expansion of $(\mathcal{M}, R)$ along $\iota_j : (\Sigma, Nom, \Lambda) \to (\Sigma, Nom \cup \{j\}, \Lambda)$ such that $j_R = s$.

$\lambda(k_1, \ldots, k_n)$ is introduced in this paper but a semantically equivalent sentence can be obtained by combining the remaining sentence operators. However, in certain fragments of hybrid logics the sentence operators are restricted making the present approach more useful. The sentence building operator @ is called *retrieve* since it changes the point of evaluation in the model. The sentence building operator $\downarrow$ is called *store* since it gives a name to the current state and it allows a reference to it. The global satisfaction holds when the satisfaction holds locally in all states, i.e. $(\mathcal{M}, R) \models^{\mathtt{HI}} \rho$ iff $(\mathcal{M}, R) \models^s \rho$ for all $s \in |R|$. Given a signature $\Delta \in |\mathbb{S}ig^{\mathtt{HI}}|$ and two sets of sentences $\Gamma, E \in \mathbb{S}en^{\mathtt{HI}}(\Delta)$, we write $\Gamma \models^{\mathtt{HI}} E$ iff for all models $(\mathcal{M}, R) \in |\mathbb{M}od^{\mathtt{HI}}(\Delta)|$ such that $(\mathcal{M}, R) \models^{\mathtt{HI}} \Gamma$ we have $(\mathcal{M}, R) \models^{\mathtt{HI}} E$. Note that variables may be interpreted differently across distinct worlds, which amounts to the world-line semantics of [14].

Satisfaction Condition. The satisfaction condition for hybrid institutions is a direct consequence of the following local satisfaction condition.

Theorem 2 [6]. *Let $\Delta = (\Sigma, Nom, \Lambda)$ and $\Delta' = (\mathbb{S}ig', Nom', \Lambda')$ be two* HI *signatures and $\varphi : \Delta \to \Delta'$ a signature morphism. For any $\rho \in \mathbb{S}en^{\mathtt{HI}}(\Delta)$, $(\mathcal{M}, R') \in \mathbb{M}od^{\mathtt{HI}}(\Delta')$ and $s \in |R'|$ we have*

$$\mathbb{M}od^{\mathtt{HI}}(\varphi)(\mathcal{M}', R') \models^s \rho \text{ iff } (\mathcal{M}', R') \models^s \mathbb{S}en^{\mathtt{HI}}(\varphi)(\rho)$$

The result of the hybridisation process is an institution.

Corollary 1 [6]. $\mathtt{HI} = (\mathbb{S}ig^{\mathtt{HI}}, \mathbb{S}en^{\mathtt{HI}}, \mathbb{M}od^{\mathtt{HI}}, \models^{\mathtt{HI}})$ *is an institution.*

A myriad of examples of hybrid institutions may be generated by applying the construction described above to various parameters: (1) the base institution I together with the quantification category $\mathcal{Q}$, and (2) by considering different constrained model functors ($\mathbb{M}od^{\mathtt{CHI}} : \mathbb{S}ig^{\mathtt{CHI}} \to \mathbb{CAT}$) for HI.

*Example 8 (Hybrid first-order logic (***HFOL***)).* This institution is obtained by applying the hybridisation process to **FOL** with the quantification subcategory consisting of signature extensions with a finite number of variables.

*Example 9 (Hybrid Propositional Logic (***HPL***)).* This institution is obtained by applying the hybridisation process to **PL** with the quantification category consisting only of identity signature morphisms. In applications, the category $\mathbb{S}ig^{\mathbf{HPL}}$ is restricted to the *full subcategory*[4] $\mathbb{S}ig^{\mathbf{HPL}'}$ which consists of

[4] A category $\mathcal{C}$ is a full subcategory of $\mathcal{C}'$ if $\mathcal{C}$ is a subcategory of $\mathcal{C}'$ and for all objects $A, B \in |C|$ we have $\mathcal{C}(A, B) = \mathcal{C}'(A, B)$.

signatures (P, Nom, Λ), where P is a set of propositional variables, Nom is a set of nominals and Λ is the family of modalities such that $\Lambda_2 = \{\lambda\}$ and $\Lambda_n = \emptyset$ for all $n \neq 2$. In this case we denote $[\Lambda]$ simply by $\Box$. Let $\mathbf{HPL}' = (\mathbb{S}ig^{\mathbf{HPL}'}, \mathbb{S}en^{\mathbf{HPL}}, \mathbb{M}od^{\mathbf{HPL}}, \models^{\mathbf{HPL}})$.

Example 10 (Constrained Hybridisation). Let $\mathtt{I} = (\mathbb{S}ig, \mathbb{S}en, \mathbb{M}od, \models)$ be a base institution, and $\mathbb{M}od^{\mathtt{CHI}} : \mathbb{S}ig^{\mathtt{CHI}} \to \mathbb{CAT}$ a constrained model functor for $\mathtt{HI}$. The *constrained hybridised institution* $\mathtt{CHI} = (\mathbb{S}ig^{\mathtt{CHI}}, \mathbb{S}en^{\mathtt{CHI}}, \mathbb{M}od^{\mathtt{CHI}}, \models^{\mathtt{CHI}})$ is obtained similarly to the case of base institutions:

(a) $\mathbb{S}en^{\mathtt{CHI}} : \mathbb{S}ig^{\mathtt{CHI}} \to \mathbb{S}et$ is the restriction of $\mathbb{S}en^{\mathtt{HI}} : \mathbb{S}ig^{\mathtt{HI}} \to \mathbb{S}et$ to $\mathbb{S}ig^{\mathtt{CHI}}$,
(b) for each signature $\Delta \in |\mathbb{S}ig^{\mathtt{CHI}}|$ and model $(\mathcal{M}, R) \in |\mathbb{M}od^{\mathtt{CHI}}(\Delta)|$,

$$(\mathcal{M}, R) \models^{\mathtt{CHI}} \rho \text{ iff } (\mathcal{M}, R) \models^{s} \rho \text{ for all } s \in |R|.$$

Note that $(\mathcal{M}, R) \models^{\mathtt{HI}} \rho$ iff $(\mathcal{M}, R) \models^{\mathtt{CHI}} \rho$. Given a signature $\Delta \in |\mathbb{S}ig^{\mathtt{CHI}}|$ and two sets of sentences $\Gamma, E \in \mathbb{S}en^{\mathtt{CHI}}(\Delta)$, we write $\Gamma \models^{\mathtt{CHI}} E$ iff for each model $(\mathcal{M}, R) \in |\mathbb{M}od^{\mathtt{CHI}}(\Delta)|$ such that $(\mathcal{M}, R) \models^{\mathtt{CHI}} \Gamma$ we have $(\mathcal{M}, R) \models^{\mathtt{CHI}} E$.

Remark 1. $\Gamma \models^{\mathtt{HI}} E$ implies $\Gamma \models^{\mathtt{CHI}} E$ but the converse implication may not hold.

Example 11 (Injective Hybridisation). Let $\mathtt{I} = (\mathbb{S}ig^{\mathtt{I}}, \mathbb{S}en^{\mathtt{I}}, \mathbb{M}od^{\mathtt{I}}, \models^{\mathtt{I}})$ be a base institution. The *injective hybridisation* $\mathtt{IHI} = (\mathbb{S}ig^{\mathtt{IHI}}, \mathbb{S}en^{\mathtt{IHI}}, \mathbb{M}od^{\mathtt{IHI}}, \models^{\mathtt{IHI}})$ of the base institution $\mathtt{I}$ is a constrained hybridised institution obtained from $\mathtt{HI} = (\mathbb{S}ig^{\mathtt{HI}}, \mathbb{S}en^{\mathtt{HI}}, \mathbb{M}od^{\mathtt{HI}}, \models^{\mathtt{HI}})$ and its constrained model functor $\mathbb{M}od^{\mathtt{IHI}} : \mathbb{S}ig^{\mathtt{IHI}} \to \mathbb{CAT}$ that do not allow confusion among nominals: (a) $\mathbb{S}ig^{\mathtt{IHI}}$ is the broad subcategory of $\mathbb{S}ig^{\mathtt{HI}}$ consisting of signature morphisms injective on nominals, i.e. φ_{Nom} is injective for all $\varphi \in \mathbb{S}ig^{\mathtt{IHI}}$, and (b) $\mathbb{M}od^{\mathtt{IHI}}(\Sigma, Nom, \Lambda)$ is the full subcategory of $\mathbb{M}od^{\mathtt{HI}}(\Sigma, Nom, \Lambda)$ consisting of models that do not allow confusion among nominals, i.e. $j_R = k_R$ implies $j = k$ for all $j, k \in Nom$.

Our results are not applicable directly to hybrid institutions but rather to their restriction to models that do not allow confusion among nominals. The following results can be instantiated, for example, to the injective hybridisation of **FOL**. However, when the quantification subcategory $\mathcal{Q}$ consists of identities (take for example **PL**) then the semantic restriction of the hybridised logic is no longer required. This means that the following results are applicable to **HPL**.

Example 12 (Quantifier-free Injective Hybridisation). The quantifier-free injective hybridisation $\mathtt{QIHI} = (\mathbb{S}ig^{\mathtt{IHI}}, \mathbb{S}en^{\mathtt{QIHI}}, \mathbb{M}od^{\mathtt{IHI}}, \models^{\mathtt{IHI}})$ of a base institution $\mathtt{I} = (\mathbb{S}ig^{\mathtt{I}}, \mathbb{S}en^{\mathtt{I}}, \mathbb{M}od^{\mathtt{I}}, \models^{\mathtt{I}})$ is obtained from the injective hybridisation $\mathtt{IHI} = (\mathbb{S}ig^{\mathtt{IHI}}, \mathbb{S}en^{\mathtt{IHI}}, \mathbb{M}od^{\mathtt{IHI}}, \models^{\mathtt{IHI}})$ by restricting the syntax to quantifier-free sentences, i.e. for each $(\Sigma, Nom, \Lambda) \in |\mathbb{S}ig^{\mathtt{IHI}}|$ the set $\mathbb{S}en^{\mathtt{QIHI}}(\Sigma, Nom, \Lambda)$ consists of sentences obtained from nominal sentences (e.g. $k \in Nom$), hybrid relational atoms (e.g. $\lambda(k_1, \ldots, k_n) \in \mathbb{S}en^{\mathtt{IHI}}(\Sigma, Nom, \Lambda)$) and the sentences in $\mathbb{S}en(\Sigma)$ by applying Boolean connectives and the operator @. This institution is useful for defining hybrid substitutions that do not involve any form of quantification (see Sect. 3.1).

3.1 Hybrid Substitutions

We extend the notion of substitution from a base institution to its hybridisation. In this subsection we assume a base institution $\mathtt{I} = (\mathbb{S}ig, \mathbb{S}en, \mathbb{M}od, \models)$, a broad subcategory of signature morphisms $\mathcal{D} \subseteq \mathbb{S}ig$, and a substitution functor $\mathbb{S}ub : \mathcal{D}^{op} \to \mathbb{CAT}$ for the base institution $\mathtt{I}$. Let $\mathcal{D}^{\mathtt{HI}} \subseteq \mathbb{S}ig^{\mathtt{HI}}$ be the broad subcategory of hybrid signature morphisms of the form $\varphi : (\Sigma, Nom, \Lambda) \to (\Sigma_1, Nom, \Lambda)$ such that $\Sigma \overset{\varphi_{\mathbb{S}ig}}{\to} \Sigma_1 \in \mathcal{D}$, $\varphi_{\mathbb{N}om} = 1_{Nom}$ and $\varphi_{\mathbb{R}el} = 1_{\Lambda}$.

Inherited Substitutions. Hybrid substitutions can be obtained from combinations of substitutions in the base institution. Let $(\Sigma, Nom, \Lambda) \in |\mathbb{S}ig^{\mathtt{IHI}}|$ be a signature and $\Theta = \{\theta_k : (\Sigma \overset{\varphi_1}{\to} \Sigma_1) \to (\Sigma \overset{\varphi_2}{\to} \Sigma_2)\}_{k \in Nom}$ a family of substitutions in $\mathbb{S}ub$. On the syntactic side, Θ determines a function

$$\Theta^k : \mathbb{S}en^{\mathtt{QIHI}}(\Sigma_1, Nom, \Lambda) \to \mathbb{S}en^{\mathtt{QIHI}}(\Sigma_2, Nom, \Lambda)$$

for each nominal $k \in Nom$:

- $\Theta^k(j) = j$, for all $j \in Nom$;
- $\Theta^k(\lambda(k_1, \ldots, k_n)) = \lambda(k_1, \ldots, k_n)$ for all $\lambda \in \Lambda_{n+1}$ and $k_i \in Nom$;
- $\Theta^k(\rho) = \theta_k(\rho)$ for any $\rho \in \mathbb{S}en^{\mathtt{I}}(\Sigma)$;
- $\Theta^k(\rho \star \rho') = \Theta^k(\rho) \star \Theta^k(\rho')$, $\star \in \{\wedge, \Rightarrow\}$;
- $\Theta^k(\neg\rho) = \neg\Theta^k(\rho)$;
- $\Theta^k(@_j \rho) = \Theta^j(\rho)$;

Since $\mathbb{S}en^{\mathtt{I}}(\varphi_1); \mathbb{S}en(\theta_k) = \mathbb{S}en^{\mathtt{I}}(\varphi_2)$ for all nominals $k \in Nom$, the following result holds.

Lemma 2. *The diagram below is commutative*

$$\begin{array}{ccc} \mathbb{S}en^{\mathtt{QIHI}}(\Sigma_1, Nom, \Lambda) & \xrightarrow{\Theta^k} & \mathbb{S}en^{\mathtt{QIHI}}(\Sigma_2, Nom, \Lambda) \\ {\scriptstyle \mathbb{S}en^{\mathtt{QIHI}}(\varphi_1)} \nwarrow & & \nearrow {\scriptstyle \mathbb{S}en^{\mathtt{QIHI}}(\varphi_2)} \\ & \mathbb{S}en^{\mathtt{QIHI}}(\Sigma, Nom, \Lambda) & \end{array}$$

for all nominals $k \in Nom$.

On the semantic side, Θ determines a functor

$$\mathbb{M}od^{\mathtt{IHI}}(\Theta^k) : \mathbb{M}od^{\mathtt{IHI}}(\Sigma_2, Nom, \Lambda) \to \mathbb{M}od^{\mathtt{IHI}}(\Sigma_1, Nom, \Lambda)$$

often denoted by $\_\upharpoonright_{\Theta^k}$ for all nominals $k \in Nom$:

- for every $(\mathcal{M}^2, R) \in |\mathbb{M}od^{\mathtt{IHI}}(\Sigma_2, Nom, \Lambda)|$, $(\mathcal{M}^2, R) \upharpoonright_{\Theta^k} = (\mathcal{M}^2 \upharpoonright_{\Theta^k}, R)$, where $\mathcal{M}^2 \upharpoonright_{\Theta^k}$ is defined by
  - $(\mathcal{M}^2 \upharpoonright_{\Theta^k})_{j_R} = \mathcal{M}^2_{j_R} \upharpoonright_{\theta_j}$ for all nominals $j \in Nom$, and
  - $(\mathcal{M}^2 \upharpoonright_{\Theta^k})_s = \mathcal{M}^2_s \upharpoonright_{\theta_k}$ for all $s \in (|R| - Nom_R)$.
- for every $h^2 : (\mathcal{M}^2, R) \to (\mathcal{N}^2, P) \in \mathbb{M}od^{\mathtt{IHI}}(\Sigma_2, Nom, \Lambda)$ we have

- $(h^2 \upharpoonright_{\Theta^k})_{j_R} = h^2_{j_R} \upharpoonright_{\theta_j}$ for all nominals $j \in Nom$, and
- $(h^2 \upharpoonright_{\Theta^k})_s = h^2_s \upharpoonright_{\theta_k}$ for all $s \in (|R| - Nom_R)$.

The definition of $\_\upharpoonright_{\Theta^k}$ is consistent because no confusion of nominals is allowed inside of the models $(\mathcal{M}^2, R)$. Since $\mathbb{M}od(\theta_k); \mathbb{M}od^{\mathtt{I}}(\varphi_1) = \mathbb{M}od^{\mathtt{I}}(\varphi_2)$ for all nominals $k \in Nom$, the following result holds.

Lemma 3. *The diagram below is commutative*

$$\begin{array}{ccc} \mathbb{M}od^{\mathtt{IHI}}(\Sigma_1, Nom, \Lambda) & \xleftarrow{\ \_\upharpoonright_{\Theta^k}\ } & \mathbb{M}od^{\mathtt{IHI}}(\Sigma_2, Nom, \Lambda) \\ & {\scriptstyle \mathbb{M}od^{\mathtt{IHI}}(\varphi_1)} \searrow \quad \swarrow {\scriptstyle \mathbb{M}od^{\mathtt{IHI}}(\varphi_2)} & \\ & \mathbb{M}od^{\mathtt{IHI}}(\Sigma, Nom, \Lambda) & \end{array}$$

for all nominals $k \in Nom$.

Next result can be regarded as the satisfaction condition for the substitutions inherited from the base institution.

Proposition 2 (Satisfaction Condition). *Given a signature $(\Sigma, Nom, \Lambda) \in |\mathbb{S}ig^{\mathtt{HI}}|$, for every model $(\mathcal{M}^2, R) \in \mathbb{M}od^{\mathtt{IHI}}(\Sigma_2, Nom, \Lambda)$ and each sentence $\rho \in \mathbb{S}en^{\mathtt{QIHI}}(\Sigma, Nom, \Lambda)$*

$$(\mathcal{M}^2, R) \models^{k_R} \Theta^k(\rho) \textit{ iff } (\mathcal{M}^2, R)\upharpoonright_{\Theta^j} \models^{k_R} \rho$$

for all nominals $j, k \in Nom$.

Proposition 2 stands at the basis of proving initiality and Herbrand's theorem in hybrid logics where the variables may be interpreted differently across distinct worlds. The following is a corollary of Proposition 2 which allows one to infer new sentences from initial axioms by applying substitutions inherited from the base institution.

Corollary 2. *Assume a signature $\Delta = (\Sigma, Nom, \Lambda) \in |\mathbb{S}ig^{\mathtt{IHI}}|$ and a hybrid substitution $\Theta = \{\theta_j : (\Sigma \overset{\varphi_1}{\to} \Sigma_1) \to (\Sigma \overset{\varphi_2}{\to} \Sigma_2)\}_{j \in Nom}$. For all sentences $\rho \in \mathbb{S}en^{\mathtt{QIHI}}(\Delta)$ ρ we have*

$$(\forall \varphi_1)\rho \models^{\mathtt{IHI}} @_k(\forall \varphi_2)\Theta^k(\rho)$$

for all nominals $k \in Nom$.

Nominal Substitutions. Nominal substitutions are captured by the notion of signature morphisms in the hybridised institution. Let $\iota_j : (\Sigma, Nom, \Lambda) \hookrightarrow (\Sigma, Nom \cup \{j\}, \Lambda)$ be a signature extension with the nominal variable j. A nominal substitution is represented by a function $\varphi_{\mathrm{Nom}} : \{j\} \to Nom$ which can be canonically extended to a signature morphism $\varphi : (\Sigma, Nom \cup \{j\}, \Lambda) \to (\Sigma, Nom, \Lambda)$. The following result is a consequence of the satisfaction condition for the hybridised institution.

Lemma 4. *Let* $(\forall j)\rho \in \mathbb{S}en^{\mathtt{HI}}(\Sigma, Nom, \Lambda)$ *and* $k \in Nom$.

(1) $(\forall j)\rho \models^{\mathtt{HI}} \rho[j \leftarrow k]$*; moreover,* $(\mathcal{M}, R) \models^{s} (\forall j)\rho$ *implies* $(M, R) \models^{s} \rho[j \leftarrow k]$ *for all models* $(\mathcal{M}, R) \in |\mathbb{M}od^{\mathtt{HI}}(\Sigma, Nom, \Lambda)|$ *and states* $s \in |R|$*;*
(2) $(\downarrow j)\rho \models^{\mathtt{HI}} @_k \rho[j \leftarrow k]$.[5]

3.2 Reachable Hybrid Models

In this subsection we extend the notion of reachability to hybrid institutions. Let $\mathtt{I} = (\mathbb{S}ig, \mathbb{S}en, \mathbb{M}od, \models)$ be a base institution, $\mathcal{D} \subseteq \mathbb{S}ig$ a broad subcategory of signature morphisms, and $\mathbb{S}ub : \mathcal{D}^{op} \to \mathbb{CAT}$ a substitution functor for $\mathtt{I}$.

Definition 4. *A model* $(\mathcal{M}, R) \in |\mathbb{M}od^{\mathtt{HI}}(\Sigma, Nom, \Lambda)|$*, where* $(\Sigma, Nom, \Lambda) \in |\mathbb{S}ig^{\mathtt{HI}}|$*, is* $\mathbb{S}ub$*-reachable if (a)* $|R| = Nom_R$*, where* $Nom_R = \{k_R \mid k \in Nom\}$*, and (b)* $\mathcal{M}_{k_R}$ *is* $\mathbb{S}ub$*-reachable in* $\mathtt{I}$ *for all nominals* $k \in Nom$.

In the injective hybridisation, the expansions of reachable models along signature morphisms in $\mathcal{D}^{\mathtt{HI}}$ generate hybrid substitutions.

Proposition 3. *Given a signature* $(\Sigma, Nom, \Lambda) \in |\mathbb{S}ig^{\mathtt{HI}}|$ *and a* $\mathbb{S}ub$*-reachable model* $(\mathcal{M}, R) \in |\mathbb{M}od^{\mathtt{IHI}}(\Sigma, Nom, \Lambda)|$ *then for every signature morphism* $\chi : (\Sigma, Nom, \Lambda) \to (\Sigma', Nom, \Lambda)$ *with* $\Sigma \xrightarrow{\chi} \Sigma' \in \mathcal{D}$ *and each* χ*-expansion* $(\mathcal{M}', R)$ *of* $(\mathcal{M}, R)$ *there exists a hybrid substitution* $\Theta = \{\chi \xrightarrow{\theta_k} 1_\Sigma\}_{k \in Nom}$ *such that* $(\mathcal{M}, R)\restriction_{\Theta^j} = (\mathcal{M}', R)$ *for all nominals* $j \in Nom$.

This definition of reachability is used in the context of injective hybridisations and their constrained sub-institutions.

4 Initiality

The following results on the existence of initial models depend on multiple parameters that can be instantiated in the same context in many ways producing different results. We will focus largely on parameter instantiation of the abstract theorems to concrete hybrid logical systems to obtain the desired applications. However, the interested reader may find other useful applications as well. In this section we assume a base institution $\mathtt{I} = (\mathbb{S}ig^{\mathtt{I}}, \mathbb{S}en^{\mathtt{I}}, \mathbb{M}od^{\mathtt{I}}, \models^{\mathtt{I}})$, a broad subcategory $\mathcal{D} \subseteq \mathbb{S}ig$ of signature morphisms and a substitution functor $\mathbb{S}ub : \mathcal{D}^{op} \to \mathbb{CAT}$ for the base institution $\mathtt{I}$.

4.1 Basic Hybrid Sentences

In addition to the assumptions made at the beginning of this section, let us consider a sub-functor $(\mathbb{S}en_0^{\mathtt{I}} : \mathbb{S}ig \to \mathbb{S}et)$ of $\mathbb{S}en^{\mathtt{I}}$. We define the sentence functor $(\mathbb{S}en_0^{\mathtt{HI}} : \mathbb{S}ig^{\mathtt{HI}} \to \mathbb{S}et)$ of $\mathbb{S}en^{\mathtt{HI}}$ for each signature $(\Sigma, Nom, \Lambda) \in |\mathbb{S}ig^{\mathtt{HI}}|$,

[5] We denote by $\rho[j \leftarrow k]$ the sentence $\varphi(\rho)$, where the signature morphism $\varphi : (\Sigma, Nom \cup \{j\}, \Lambda) \to (\Sigma, Nom, \Lambda)$ is the canonical extension of the function $\varphi_{\mathsf{Nom}} : \{j\} \to Nom$ defined by $\varphi_{\mathsf{Nom}}(j) = k$.

(1) $@_j k \in \mathbb{S}en_0^{\mathtt{HI}}(\Sigma, Nom, \Lambda)$ for all $j, k \in Nom$,
(2) $@_j \lambda(k_1, \dots, k_n) \in \mathbb{S}en_0^{\mathtt{HI}}(\Sigma, Nom, \Lambda)$ for all $\lambda \in \Lambda_{n+1}$, $j \in Nom$ and $k_i \in Nom$,
(3) $@_j \rho \in \mathbb{S}en_0^{\mathtt{HI}}(\Sigma, Nom, \Lambda)$ for all $j \in Nom$ and $\rho \in \mathbb{S}en_0^{\mathtt{I}}(\Sigma)$.

In concrete examples of institutions, $\mathtt{I}_0 = (\mathbb{S}ig^{\mathtt{I}}, \mathbb{S}en_0^{\mathtt{I}}, \mathbb{M}od^{\mathtt{I}}, \models^{\mathtt{I}})$ is the restriction of the base institution to atomic sentences, and the institution $\mathtt{HI}_0 = (\mathbb{S}ig^{\mathtt{HI}}, \mathbb{S}en_0^{\mathtt{HI}}, \mathbb{M}od^{\mathtt{HI}}, \models^{\mathtt{HI}})$ gives the building bricks for constructing theories that have initial models in the hybridised institution.

Theorem 3. *If every set of sentences of* $\mathtt{I}_0$ *is epi basic then every set of sentences of* $\mathtt{HI}_0$ *is epi basic. Moreover, if each set of sentences of* $\mathtt{I}_0$ *has a basic model that is* $\mathbb{S}ub$*-reachable then each set of sentences of* $\mathtt{HI}_0$ *has a basic model that is* $\mathbb{S}ub$*-reachable.*

We apply Theorem 3 to **HFOL**. Let $\mathbb{S}en_0^{\mathbf{FOL}} : \mathbb{S}ig^{\mathbf{FOL}} \to \mathbb{S}et$ be the sub-functor of $\mathbb{S}en^{\mathbf{FOL}}$ such that for any signature $\Sigma \in |\mathbb{S}ig^{\mathbf{FOL}}|$ the set $\mathbb{S}en_0^{\mathbf{FOL}}(\Sigma)$ consists of atoms. We define $\mathbf{FOL}_0 = (\mathbb{S}ig^{\mathbf{FOL}}, \mathbb{S}en_0^{\mathbf{FOL}}, \mathbb{M}od^{\mathbf{FOL}}, \models^{\mathbf{FOL}})$ and $\mathbf{HFOL}_0 = (\mathbb{S}ig^{\mathbf{HFOL}}, \mathbb{S}en_0^{\mathbf{HFOL}}, \mathbb{M}od^{\mathbf{HFOL}}, \models^{\mathbf{HFOL}})$ using the general pattern described above.

Corollary 3. *All sets of* $\mathbf{HFOL}_0$ *sentences are epi basic and the corresponding basic models are* $\mathbb{S}ub^{\mathbf{FOL}}$*-reachable.*

Proof. By Lemma 1, any set of **FOL** atoms is epi basic. By Proposition 1, the corresponding basic models are $\mathbb{S}ub^{\mathbf{FOL}}$-reachable. By Theorem 3, any set of $\mathbf{HFOL}_0$ sentences is epi basic and the corresponding basic models are $\mathbb{S}ub^{\mathbf{FOL}}$-reachable. □

We return to the general setting and we define the sub-functor $(\mathbb{S}en_0^{\mathtt{IHI}} : \mathbb{S}ig^{\mathtt{IHI}} \to \mathbb{S}et)$ of $\mathbb{S}en^{\mathtt{IHI}}$ for each signature $(\Sigma, Nom, \Lambda) \in |\mathbb{S}ig^{\mathtt{HI}}|$,

(1) $@_j \lambda(k_1, \dots, k_n) \in \mathbb{S}en_0^{\mathtt{IHI}}(\Sigma, Nom, \Lambda)$ for any $\lambda \in \Lambda_{n+1}$, $j \in Nom$ and $k_i \in Nom$,
(2) $@_j \rho \in \mathbb{S}en_0^{\mathtt{IHI}}(\Sigma, Nom, \Lambda)$ for any $j \in Nom$ and $\rho \in \mathbb{S}en_0^{\mathtt{I}}(\Sigma)$.

The institution $\mathtt{IHI}_0 = (\mathbb{S}ig^{\mathtt{IHI}}, \mathbb{S}en_0^{\mathtt{IHI}}, \mathbb{M}od^{\mathtt{IHI}}, \models^{\mathtt{IHI}})$ gives the building bricks for constructing theories that have initial models in the injective hybridisation.

Theorem 4. *If every set of sentences of* $\mathtt{I}_0$ *is epi basic then every set of sentences of* $\mathtt{IHI}_0$ *is epi basic. Moreover, if each set of sentences of* $\mathtt{I}_0$ *has a basic model that is* $\mathbb{S}ub$*-reachable then each set of sentences of* $\mathtt{IHI}_0$ *has a basic model that is* $\mathbb{S}ub$*-reachable.*

We apply Theorem 4 to **IHFOL**. Using the general pattern described above, let us define $\mathbf{IHFOL}_0 = (\mathbb{S}ig^{\mathbf{IHFOL}}, \mathbb{S}en_0^{\mathbf{IHFOL}}, \mathbb{M}od^{\mathbf{IHFOL}}, \models^{\mathbf{IHFOL}})$ as the injective hybridisation of **FOL**.

Corollary 4. *All sets of* $\mathbf{IHFOL}_0$ *sentences are epi basic and the corresponding basic models are* $\mathbb{S}ub^{\mathbf{FOL}}$*-reachable.*

Proof. By Lemma 1, any set of atoms in **FOL** is epi basic. By Proposition 1, the corresponding basic models are $\mathbb{S}ub^{\mathbf{FOL}}$-reachable. By Theorem 4, any set of sentences in $\mathbf{HFOL}_0$ is epi basic and the corresponding basic models are $\mathbb{S}ub^{\mathbf{FOL}}$-reachable. □

In the next subsections we prove that the initiality property is closed under the following sentence building operators: logical implication $\Rightarrow$, universal quantification $\forall$, store $\downarrow$, and box $\Box$.

4.2 Implication

In addition to the assumptions made at the beginning of this section, let us consider a constrained model functor $\mathbb{M}od^{\mathtt{CHI}} : \mathbb{S}ig^{\mathtt{CHI}} \to \mathbb{CAT}$ for HI, and three sub-functors $(\mathbb{S}en_*^{\mathtt{CHI}} : \mathbb{S}ig^{\mathtt{CHI}} \to \mathbb{S}et)$, $(\mathbb{S}en_\bullet^{\mathtt{CHI}} : \mathbb{S}ig^{\mathtt{CHI}} \to \mathbb{S}et)$ and $(\mathbb{S}en_1^{\mathtt{CHI}} : \mathbb{S}ig^{\mathtt{CHI}} \to \mathbb{S}et)$ of $\mathbb{S}en^{\mathtt{CHI}}$ such that any sentence of $\mathbb{S}en_1^{\mathtt{CHI}}(\Delta)$, where $\Delta \in |\mathbb{S}ig^{\mathtt{CHI}}|$, is semantically equivalent in CHI to a sentence of the form $\bigwedge H \Rightarrow C$, where $H \subseteq \mathbb{S}en_*^{\mathtt{CHI}}(\Delta)$ and $C \in \mathbb{S}en_\bullet^{\mathtt{CHI}}(\Delta)$.

Theorem 5. *If for each signature $\Delta \in |\mathbb{S}ig^{\mathtt{CHI}}|$,*

(1) any set $B \subseteq \mathbb{S}en_*^{\mathtt{CHI}}(\Delta)$ is basic in HI,[6] *and*
(2) any set $\Gamma \subseteq \mathbb{S}en_\bullet^{\mathtt{CHI}}(\Delta)$ has an initial $\mathbb{S}ub$-reachable model $(\mathcal{M}^\Gamma, R^\Gamma) \in |\mathbb{M}od^{\mathtt{CHI}}(\Delta)|$,

then any set of sentences of the institution $\mathtt{CHI}_1 = (\mathbb{S}ig^{\mathtt{CHI}}, \mathbb{S}en_1^{\mathtt{CHI}}, \mathbb{M}od^{\mathtt{CHI}}, \models^{\mathtt{CHI}})$ has an initial $\mathbb{S}ub$-reachable model.

We apply Theorem 5 to **HFOL**. The constrained model functor $\mathbb{M}od^{\mathtt{CHI}}$ is $\mathbb{M}od^{\mathbf{HFOL}} : \mathbb{S}ig^{\mathbf{HFOL}} \to \mathbb{CAT}$. The functors $\mathbb{S}en_*^{\mathtt{CHI}}$ and $\mathbb{S}en_\bullet^{\mathtt{CHI}}$ are both instantiated to $\mathbb{S}en_0^{\mathbf{HFOL}} : \mathbb{S}ig^{\mathbf{HFOL}} \to \mathbb{S}et$. The institution $\mathtt{CHI}_1$ is $\mathbf{HFOL}_1$, the restriction of **HFOL** to sentences of the form $\bigwedge H \Rightarrow C$, where $H \cup \{C\}$ is a set of $\mathbf{HFOL}_0$ sentences.

Corollary 5. *Any set of* $\mathbf{HFOL}_1$ *sentences has an initial $\mathbb{S}ub^{\mathbf{FOL}}$-reachable model.*

We apply Theorem 5 to **IHFOL**. The institution CHI is **IHFOL**. The functor $\mathbb{S}en_*^{\mathtt{CHI}}$ is the restriction of $(\mathbb{S}en_0^{\mathbf{HFOL}} : \mathbb{S}ig^{\mathbf{HFOL}} \to \mathbb{S}et)$ to $\mathbb{S}ig^{\mathbf{IHFOL}}$. The functor $\mathbb{S}en_\bullet^{\mathtt{CHI}}$ is $(\mathbb{S}en_0^{\mathbf{IHFOL}} : \mathbb{S}ig^{\mathbf{IHFOL}} \to \mathbb{S}et)$. The sentence functor $\mathbb{S}en_1^{\mathtt{CHI}}$ is $(\mathbb{S}en_1^{\mathbf{IHFOL}} : \mathbb{S}ig^{\mathbf{IHFOL}} \to \mathbb{S}et)$ such that for all $\Delta \in |\mathbb{S}ig^{\mathbf{IHFOL}}|$ the set $\mathbb{S}en_1^{\mathbf{IHFOL}}(\Delta)$ consists of sentences of the form $\bigwedge H \Rightarrow C$, where $H \subseteq \mathbb{S}en_0^{\mathbf{HFOL}}(\Delta)$ and $C \in \mathbb{S}en_0^{\mathbf{IHFOL}}(\Delta)$.

Corollary 6. *Any set of* $\mathbf{IHFOL}_1$ *sentences has an initial $\mathbb{S}ub^{\mathbf{FOL}}$-reachable model, where $\mathbf{IHFOL}_1 = (\mathbb{S}ig^{\mathbf{IHFOL}}, \mathbb{S}en_1^{\mathbf{IHFOL}}, \mathbb{M}od^{\mathbf{IHFOL}}, \models^{\mathbf{IHFOL}})$.*

[6] This condition implies that there exists a basic model $(\mathcal{M}^B, R^B) \in |\mathbb{M}od^{\mathtt{HI}}(\Delta)|$, but it is also possible that $(\mathcal{M}^B, R^B) \notin |\mathbb{M}od^{\mathtt{CHI}}(\Delta)|$.

4.3 Nominal Quantification

In addition to the assumptions made at the beginning of this section, let us consider a constrained model functor $\mathbb{M}od^{\mathtt{CHI}} : \mathbb{S}ig^{\mathtt{CHI}} \to \mathbb{CAT}$ for $\mathtt{HI}$, and two sub-functors $(\mathbb{S}en_1^{\mathtt{CHI}} : \mathbb{S}ig^{\mathtt{CHI}} \to \mathbb{S}et)$ and $(\mathbb{S}en_2^{\mathtt{CHI}} : \mathbb{S}ig^{\mathtt{CHI}} \to \mathbb{S}et)$ of $\mathbb{S}en^{\mathtt{CHI}}$ such that all sentences of $\mathtt{CHI}_2 = (\mathbb{S}ig^{\mathtt{CHI}}, \mathbb{S}en_2^{\mathtt{CHI}}, \mathbb{M}od^{\mathtt{CHI}}, \models^{\mathtt{CHI}})$ are semantically equivalent in $\mathtt{CHI}$ to a sentence of the form $(\forall j)\rho$, where j is a nominal variable and ρ is a sentence of $\mathtt{CHI}_1 = (\mathbb{S}ig^{\mathtt{CHI}}, \mathbb{S}en_1^{\mathtt{CHI}}, \mathbb{M}od^{\mathtt{CHI}}, \models^{\mathtt{CHI}})$.

Theorem 6. *If every set of sentences of* $\mathtt{CHI}_1$ *has an initial* $\mathbb{S}ub$*-reachable model then each set of sentences of* $\mathtt{CHI}_2$ *has an initial* $\mathbb{S}ub$*-reachable model.*

The following result is essential for applying Theorem 6 to concrete examples of institutions.

Lemma 5. *In the institution* $\mathtt{CHI}$*, any sentence* $\bigwedge H \Rightarrow C$ *is semantically equivalent to* $(\forall j) \bigwedge\{@_j h \mid h \in H\} \Rightarrow @_j C$*, and any sentence* $@_j @_k \rho$ *is semantically equivalent to* $@_k \rho$.

We apply Theorem 6 on top of $\mathbf{HFOL}_1$ defined in Subsect. 4.2. The institution $\mathtt{CHI}$ is $\mathbf{HFOL}$, and the institution $\mathtt{CHI}_1$ is $\mathbf{HFOL}_1$. The sentence functor $\mathbb{S}en_2^{\mathtt{CHI}}$ is $(\mathbb{S}en_2^{\mathbf{HFOL}} : \mathbb{S}ig^{\mathbf{HFOL}} \to \mathbb{S}et)$ which associates to each signature $\Delta = (\Sigma, Nom, \Lambda) \in |\mathbb{S}ig^{\mathbf{HFOL}}|$ the set of sentences of the form $\bigwedge H \Rightarrow C$, where $H \cup \{C\}$ consists of sentences obtained from nominal sentences (e.g. $k \in Nom$), hybrid relational atoms (e.g. $\lambda(k_1, \ldots, k_n) \in \mathbb{S}en^{\mathbf{IHFOL}}(\Delta)$) and **FOL** atoms (e.g. $t_1 = t_2 \in \mathbb{S}en_0^{\mathbf{FOL}}(\Sigma)$ and $\pi(t_1, \ldots, t_n) \in \mathbb{S}en_0^{\mathbf{FOL}}(\Sigma)$) by applying the sentence building operator @.[7]

Corollary 7. *Any set of sentences in* $\mathbf{HFOL}_2$ *has an initial* $\mathbb{S}ub^{\mathbf{FOL}}$*-reachable model.*

Proof. By Lemma 5, any sentence in $\mathbf{HFOL}_2$ is semantically equivalent to a sentence of the form $(\forall j)\rho$, where j is a nominal variable and ρ is a sentence of $\mathbf{HFOL}_1$. By Corollary 6, any set of sentences in $\mathbf{HFOL}_1$ has an initial $\mathbb{S}ub^{\mathbf{FOL}}$-reachable model. By Theorem 6, any set of sentences in $\mathbf{HFOL}_2$ has an initial $\mathbb{S}ub^{\mathbf{FOL}}$-reachable model. □

We apply Theorem 6 on top of $\mathbf{HFOL}_2$. Let $\mathbf{HFOL}_3$ be the institution obtained from **HFOL** by restricting the syntax to sentences of the form $(\forall J)\rho$, where J is a finite set of nominal variables and ρ is a quantifier-free sentence of $\mathbf{HFOL}_2$.

Corollary 8. *Any set of* $\mathbf{HFOL}_3$ *sentences has an initial* $\mathbb{S}ub^{\mathbf{FOL}}$*-reachable model.*

We call the $\mathbf{HFOL}_3$ sentences *hybrid Horn clauses of the institution* **HFOL**. Since **PL** is obtained from **FOL** by restricting the category of signatures, Corollary 8 holds also for **HPL**. We apply Theorem 6 on top of $\mathbf{IHFOL}_1$ defined

[7] The institution $\mathbf{HFOL}_2$ contains also sentences that are free of @. It follows that $\mathbb{S}en_1^{\mathbf{HFOL}}(\Delta) \subsetneq \mathbb{S}en_2^{\mathbf{HFOL}}(\Delta)$ for all $\Delta \in |\mathbb{S}ig^{\mathbf{HFOL}}|$.

in Subsect. 4.2. The sentence functor $\mathbb{S}en_1^{\mathtt{CHI}}$ is $(\mathbb{S}en_1^{\mathbf{IHFOL}} : \mathbb{S}ig^{\mathbf{IHFOL}} \to \mathbb{S}et)$. The functor $\mathbb{S}en_2^{\mathtt{CHI}}$ is $(\mathbb{S}en_2^{\mathbf{IHFOL}} : \mathbb{S}ig^{\mathbf{IHFOL}} \to \mathbb{S}et)$ which associates to each signature the set of sentences of the form $\bigwedge H \Rightarrow C$, where

(a) H consists of sentences obtained from nominal sentences, hybrid relational atoms, and **FOL** atoms by applying the sentence building operator @, and
(b) C is a sentence obtained from hybrid relational atoms and **FOL** atoms by applying @.

Corollary 9. *Any set of sentences in* $\mathbf{IHFOL}_2$ *has an initial* $\mathbb{S}ub^{\mathbf{FOL}}$*-reachable model.*

Another application of Theorem 6 can be found in Subsect. 4.4.

4.4 Inherited Quantification

In addition to the assumptions made at the beginning of this section, let us consider a constrained model functor $\mathbb{M}od^{\mathtt{CIHI}} : \mathbb{S}ig^{\mathtt{CIHI}} \to \mathbb{CAT}$ for the injective hybridisation IHI, a quantification subcategory $\mathcal{Q} \subseteq \mathcal{D}$, and two sub-functors $(\mathbb{S}en_2^{\mathtt{CIHI}} : \mathbb{S}ig^{\mathtt{CIHI}} \to \mathbb{S}et)$ and $(\mathbb{S}en_3^{\mathtt{CIHI}} : \mathbb{S}ig^{\mathtt{CIHI}} \to \mathbb{S}et)$ of $\mathbb{S}en^{\mathtt{CIHI}}$ such that

(1) the sentences of $\mathtt{CIHI}_2 = (\mathbb{S}ig^{\mathtt{CIHI}}, \mathbb{S}en_2^{\mathtt{CIHI}}, \mathbb{M}od^{\mathtt{CIHI}}, \models^{\mathtt{CIHI}})$ are semantically closed to @, i.e. for all $\Delta \in |\mathbb{S}ig^{\mathtt{CIHI}}|$, $k \in Nom$ and $\rho \in \mathbb{S}en_2^{\mathtt{CIHI}}(\Delta)$ there exists $\varepsilon \in \mathbb{S}en_2^{\mathtt{CIHI}}(\Delta)$ such that $@_k\rho \models\!\mid^{\mathtt{CIHI}} \varepsilon$,
(2) In CIHI, any sentence of $\mathbb{S}en_3(\Sigma, Nom, \Lambda)$, where $(\Sigma, Nom, \Lambda) \in |\mathbb{S}ig^{\mathtt{CIHI}}|$, is semantically equivalent to a sentence of $(\forall\chi)\rho$, where $(\Sigma, Nom, \Lambda) \xrightarrow{\chi} (\Sigma', Nom, \Lambda) \in \mathcal{Q}^{\mathtt{HI}}$ and $\rho \in \mathbb{S}en_2^{\mathtt{CIHI}}(\Sigma', Nom, \Lambda)$,
(3) for any $(\Sigma, Nom, \Lambda) \in |\mathbb{S}ig^{\mathtt{CIHI}}|$ and $\Sigma \xrightarrow{\chi} \Sigma' \in \mathcal{D}$ we have $(\Sigma', Nom, \Lambda) \in |\mathbb{S}ig^{\mathtt{CIHI}}|$, and
(4) for any hybrid substitution $\Theta = \{(\Sigma \xrightarrow{\chi_1} \Sigma_1) \xrightarrow{\theta_k} (\Sigma \xrightarrow{\chi_2} \Sigma_2)\}_{k \in Nom}$ and sentence $\rho \in \mathbb{S}en_2^{\mathtt{CIHI}}(\Sigma_1, Nom, \Lambda)$ we have $\Theta^k(\rho) \in \mathbb{S}en_2^{\mathtt{CIHI}}(\Sigma_2, Nom, \Lambda)$.

Theorem 7. *If every set of sentences of* $\mathtt{CIHI}_2$ *has an initial* $\mathbb{S}ub$*-reachable model then each set of sentences of* $\mathtt{CIHI}_3$ *has an initial* $\mathbb{S}ub$*-reachable model.*

We apply Theorem 7 on top of $\mathbf{IHFOL}_2$ defined in Subsect. 4.3. The institution CIHI is **IHFOL**, and the institution $\mathtt{CIHI}_2$ is $\mathbf{IHFOL}_2$. Note that $\mathbf{IHFOL}_2$ is closed to @, which means that assumption (1) of this subsection holds. The institution $\mathtt{CIHI}_3$ is $\mathbf{IHFOL}_3$, the restriction of **IHFOL** to sentences of the form $(\forall X)\rho$, where X is a finite set of first-order variables and ρ is a sentence in $\mathbf{IHFOL}_2$.[8] This implies that assumption (2) of this subsection holds. Since $\mathcal{D}^{\mathtt{HI}} \subseteq \mathbb{S}ig^{\mathbf{IHFOL}}$, assumption (3) of this subsection holds. All sentences of $\mathbf{IHFOL}_2$ are quantifier-free and modal-free, and by applying a hybrid substitution to a $\mathbf{IHFOL}_2$ sentence, the result is also a $\mathbf{IHFOL}_2$ sentence. It follows that assumption (4) of this subsection holds.

[8] Note that $(\forall X)\rho$ is an abbreviation for $(\forall\chi)\rho$, where $\chi : (\Sigma, Nom, \Lambda) \hookrightarrow (\Sigma[X], Nom, \Lambda) \in \mathcal{Q}^{\mathbf{HFOL}}$ is a signature extension with the finite set of first-order variables X and $\rho \in \mathbb{S}en_2^{\mathbf{IHFOL}}(\Sigma[X], Nom, \Lambda)$.

Corollary 10. *Any set of* **IHFOL$_3$** *sentences has an initial* $\mathbb{S}ub^{\mathbf{FOL}}$*-reachable model.*

We apply Theorem 6 on top of the institution **IHFOL$_3$** defined above. Let **IHFOL$_4$** be the institution obtained from **IHFOL** by restricting the syntax to sentences of the form $(\forall J)\rho$, where J is a finite set of nominal variables and ρ is a sentence of **IHFOL$_3$**.

Corollary 11. *Every set of* **IHFOL$_4$** *sentences has an initial* $\mathbb{S}ub^{\mathbf{FOL}}$*-reachable model.*

We call the **IHFOL$_4$** sentences *hybrid Horn clauses of the institution* **IHFOL**. Defining a paramodulation procedure for **IHFOL$_4$** is future research. However, the results obtained in this paper set the foundation for this direction of research.

Any sentence of the form $(\downarrow j)\rho$ is semantically equivalent to $(\forall j)j \Rightarrow \rho$. It follows that initiality is closed under store $\downarrow$. If $\lambda \in \Lambda_2$ then $[\lambda](\rho)$ is semantically equivalent to $(\forall k)\lambda(k) \Rightarrow @_k\rho$. It follows that initiality is closed under box $\Box$ when $\Lambda_n = \emptyset$ for all $n \neq 2$.

5 Herbrand's Theorem

We prove a version of Herbrand's theorem in the framework of hybrid institutions.

Theorem 8. *Let* $\mathtt{I} = (\mathbb{S}ig, \mathbb{S}en, \mathbb{M}od, \models)$ *be an institution,* $\mathcal{D} \subseteq \mathbb{S}ig$ *a broad subcategory of signature morphisms,* $\mathbb{S}ub : \mathcal{D}^{op} \to \mathbb{CAT}$ *a substitution functor for* $\mathtt{I}$ *and* $\mathcal{Q} \subseteq \mathcal{D}$ *a quantification subcategory. Consider a constrained model functor* $\mathbb{M}od^{\mathtt{CIHI}} : \mathbb{S}ig^{\mathtt{CIHI}} \to \mathbb{CAT}$ *for the injective hybridisation* IHI *such that*

(1) for any $(\Sigma, Nom, \Lambda) \in |\mathbb{S}ig^{\mathtt{CIHI}}|$ *and* $\Sigma \xrightarrow{\chi} \Sigma' \in \mathcal{D}$ *we have* $(\Sigma', Nom, \Lambda) \in |\mathbb{S}ig^{\mathtt{CIHI}}|$.

Assume a sub-functor $(\mathbb{S}en_b^{\mathtt{CIHI}} : \mathbb{S}ig^{\mathtt{CIHI}} \to \mathbb{S}et)$ *of* $\mathbb{S}en^{\mathtt{CIHI}}$ *such that*

(2) any $B \subseteq \mathbb{S}en_b^{\mathtt{CIHI}}(\Sigma, Nom, \Lambda)$ *is basic in* HI, *where* $(\Sigma, Nom, \Lambda) \in |\mathbb{S}ig^{\mathtt{CIHI}}|$.

Let $\Delta = (\Sigma, Nom, \Lambda) \in |\mathbb{S}ig^{\mathtt{CIHI}}|$ *be a signature,* $k \in Nom$ *a nominal,* $\Gamma \subseteq \mathbb{S}en^{\mathtt{CIHI}}(\Delta)$ *a set of sentences that has an initial* $\mathbb{S}ub$*-reachable model* $(\mathcal{M}^\Gamma, R^\Gamma) \in |\mathbb{M}od^{\mathtt{CIHI}}(\Delta)|$, *and* $(\exists J)(\exists \chi)\rho \in \mathbb{S}en^{\mathtt{CIHI}}(\Delta)$ *a sentence such that (a)* J *is a set of nominal variables, (b)* $\Delta \xrightarrow{\chi} \Delta' \in \mathcal{Q}^{\mathtt{HI}}$ *with* $\Delta' = (\Sigma', Nom, \Lambda)$, *and (c)* $\rho \in \mathbb{S}en_b^{\mathtt{CIHI}}(\Delta'[J])$ *with* $\Delta'[J] = (\Sigma', Nom \cup J, \Lambda)$. *Then the following statements are equivalent:*

(i) $\Gamma \models^{\mathtt{CIHI}} @_k(\exists J)(\exists \chi)\rho$,
(ii) $(\mathcal{M}^\Gamma, R^\Gamma) \models^{k_{(R^\Gamma)}} (\exists J)(\exists \chi)\rho$,
(iii) there is a hybrid substitution $\Theta = \{\theta_j : (\Sigma \xrightarrow{\chi} \Sigma') \to (\Sigma \xrightarrow{\varphi} \Sigma'')\}_{j \in Nom}$ *and a nominal substitution* $\psi : J \to Nom$ *such that* $\Gamma \models^{\mathtt{CIHI}} @_k(\forall \varphi)\Theta^k(\psi(\rho))$ *and* $\varphi : (\Sigma, Nom, \Lambda) \to (\Sigma'', Nom, \Lambda)$ *is conservative in* CIHI.

The pair of substitutions $\langle\psi, \Theta\rangle$ from the statement (iii) of Theorem 8 are called *solutions*. The sentence $@_k(\exists J)(\exists\chi)\rho$ is a *query*. The implication $(i) \Rightarrow (iii)$ reduces the satisfiability of a query by a program (represented here by a hybrid theory) to the search of a pair of substitutions, while the converse implication $(iii) \Rightarrow (i)$ shows that solutions are sound with respect to the given program. We apply Theorem 8 to **IHFOL**. Since $\mathcal{D}^{\mathbf{HFOL}} \subseteq \mathbb{S}ig^{\mathbf{IHFOL}}$ the second hypothesis holds. The functor $\mathbb{S}en_b^{\mathrm{CIHI}}$ is $(\mathbb{S}en_b^{\mathbf{IHFOL}} : \mathbb{S}ig^{\mathbf{IHFOL}} \to \mathbb{S}et)$ such that for each $(\Sigma, Nom, \Lambda) \in |\mathbb{S}ig^{\mathbf{IHFOL}}|$ the set $\mathbb{S}en_b^{\mathbf{IHFOL}}(\Sigma, Nom, \Lambda)$ consists of finite conjunctions of sentences in $\mathbb{S}en_0^{\mathbf{HFOL}}(\Sigma, Nom, \Lambda)$. By Corollary 3, any set of sentences in $\mathbf{HFOL}_0$ is epi basic in **HFOL**, which implies that any conjunction of sentences in $\mathbf{HFOL}_0$ is also epi basic in **HFOL**. It follows that condition (2) of Theorem 8 holds. By Corollary 11, any set of hybrid Horn clauses in $\mathbf{IHFOL}_4$ has an initial model. Note that for any nominal k, we have $(\exists J)(\exists X)\rho \models\!\mid^{\mathbf{HFOL}} @_k(\exists J)(\exists X)\rho$, which implies $(\exists J)(\exists X)\rho \models\!\mid^{\mathbf{IHFOL}} @_k(\exists J)(\exists X)\rho$. In **IHFOL**, the queries are sentences of the form $(\exists J)(\exists X)\rho$, where ρ is a finite conjunction of $\mathbf{HFOL}_0$ sentences.

Corollary 12. *For any set of sentences $\Gamma \subseteq \mathbb{S}en^{\mathbf{IHFOL}_4}((S, F, P), Nom, \Lambda)$, where $((S, F, P), Nom, \Lambda) \in |\mathbb{S}ig^{\mathbf{IHFOL}}|$, and any query $(\exists J)(\exists X)\rho$, where ρ is a finite conjunction of sentences in $\mathbb{S}en_0^{\mathbf{HFOL}}((S, F \cup X, P), Nom \cup J, \Lambda)$ the followings are equivalent:*

(i) $\Gamma \models^{\mathbf{IHFOL}} (\exists J)(\exists X)\rho$,
(ii) $(\mathcal{M}^\Gamma, R^\Gamma) \models^{\mathbf{IHFOL}} (\exists J)(\exists X)\rho$,
(iii) there exists a hybrid substitution $\Theta = \{\theta_j : X \to T_{(S,F,P)}(Y)\}_{j\in Nom}$ and a nominal substitution $\psi : J \to Nom$ such that $\Gamma \models^{\mathbf{IHFOL}} (\forall Y)\Theta^k(\psi(\rho))$ for some $k \in Nom$ and the sorts of variables in Y are inhabited, i.e. for any sort $s \in S$ and variable $y \in Y_s$ there exists a term $t \in T_{(S,F,P)}$.

The inhabitation requirement for the sorts of the variables in Y means that the inclusion $\iota_y : ((S, F, P), Nom, \Lambda) \to ((S, F \cup Y, P), Nom, \Lambda)$ is conservative. The restriction to injective hybridisations required by Theorem 8 is not needed if the quantification subcategory consists of identities. For example, one can prove a version of Herbrand's theorem for hybrid institutions that can be instantiated to **HPL**.

6 Conclusions

In this paper we have proved the existence of initial models of hybrid Horn clauses. Our initiality results are not based on inclusion systems and quasi-varieties as in [6]. The proof follows the structure of the sentences in the style of [9]. We assume that the atomic sentences of the base institution are epi basic and then the initiality property is proved to be closed under certain sentence building operators. This approach requires less model theoretic infrastructure than [6] and it can be applied to theories for which the corresponding class of models does not form a quasi-variety. We have developed denotational foundations for logic programming in hybrid logics independently of the details of the underlying base institution by employing institutional concepts of quantification, substitution, reachable model and basic set of sentences. In this general setting we have proved Herbrand's theorem. A future direction of research is developing a paramodulation procedure for hybrid logics. The results presented in this paper which do not involve inherited quantification can be applied to hybrid logics with model constraints [6], but much work is needed to cover the rigid quantification [2]. This constitutes another future direction of research.

References

1. Blackburn, P.: Representation, reasoning, and relational structures: a hybrid logic manifesto. Log. J. IGPL **8**(3), 339–365 (2000)
2. Braüner, T.: Hybrid Logic and its Proof-Theory. Applied Logic Series, vol. 37. Springer, Berlin (2011)
3. Diaconescu, R.: Institution-independent Ultraproducts. Fundamenta Informaticæ **55**(3–4), 321–348 (2003)
4. Diaconescu, R.: Herbrand theorems in arbitrary institutions. Inf. Process. Lett. **90**, 29–37 (2004)
5. Diaconescu, R.: Institution-independent Model Theory. Studies in Universal Logic. Birkhäuser, Basel (2008)
6. Diaconescu, R.: Quasi-varieties and initial semantics in hybridized institutions. J. Logic Comput. (2014). doi:10.1093/logcom/ext016
7. Goguen, J., Burstall, R.: Institutions: abstract model theory for specification and programming. J. Assoc. Comput. Mach. **39**(1), 95–146 (1992)
8. Goguen, J.A., Thatcher, J.W.: Initial algebra semantics. In: SWAT (FOCS), pp. 63–77. IEEE Computer Society (1974)
9. Găină, D., Futatsugi, K.: Initial semantics in logics with constructors. J. Log. Comput. **25**(1), 95–116 (2015). http://dx.doi.org/10.1093/logcom/exs044
10. Găină, D., Futatsugi, K., Ogata, K.: Constructor-based Logics. J. Univ. Comput. Sci. **18**(16), 2204–2233 (2012)
11. Găină, D., Petria, M.: Completeness by Forcing. J. Log. Comput. **20**(6), 1165–1186 (2010)
12. Lloyd, J.: Foundation of Logic Programming. Springer, Berlin (1987)
13. Martins, M.A., Madeira, A., Diaconescu, R., Barbosa, L.S.: Hybridization of institutions. In: Corradini, A., Klin, B., Cîrstea, C. (eds.) CALCO 2011. LNCS, vol. 6859, pp. 283–297. Springer, Heidelberg (2011)
14. Schurz, G.: Combinations and completeness transfer for quantified modal logics. Log. J. IGPL **19**(4), 598–616 (2011). http://dx.doi.org/10.1093/jigpal/jzp085

What Is a Derived Signature Morphism?

Till Mossakowski[1](✉), Ulf Krumnack[2], and Tom Maibaum[3]

[1] Otto-von-Guericke University of Magdeburg, Magdeburg, Germany
till@iws.cs.ovgu.de
[2] University of Osnabrück, Osnabrück, Germany
[3] McMaster University, Hamilton, Canada

Abstract. The notion of signature morphism is basic to the theory of institutions. It provides a powerful primitive for the study of specifications, their modularity and their relations in an abstract setting. The notion of *derived* signature morphism generalises signature morphisms to more complex constructions, where symbols may be mapped not only to symbols, but to arbitrary terms. The purpose of this work is to study derived signature morphisms in an institution-independent way. We will recall and generalize two known approaches to derived signature morphisms, introduce a third one, and discuss their pros and cons. We especially study the existence of colimits of derived signature morphisms. The motivation is to give an independent semantics to the notion of derived signature morphism, query and substitution in the context of the Distributed Ontology, Modeling and Specification Language DOL.

1 Introduction

The notion of signature morphism is basic to the theory of institutions. It provides a powerful primitive for the study of specifications, their modularity and their relations in an abstract setting. The notion of *derived* signature morphism generalises signature morphisms to more complex constructions, where symbols may be mapped not only to symbols, but to arbitrary terms. Derived signature morphisms have been introduced in [15] and studied in [5,6,16,20,21]. Recently, the notion of derived signature morphism has gained attention in the field of model-driven engineering [9], databases [8], analogies [23], and ontologies[1].

In this paper we investigate derived signature morphism and their properties. We recall and generalize two known approaches to derived signature morphisms, and introduce a third one. All current works define derived signature morphisms in specific institutions. We look for a way to formulate the concept in an *institution-independent* way. Especially we look for a semantics of derived signature morphisms in languages with institution-independent semantics. We also investigate the question to what extent we can *combine* systems along derived signature morphisms (via *colimits*).

---

[1] Cmp. the work on the new OMG standard. *Distributed Ontology, Modeling and Specification Language (DOL)*, see http://ontoiop.org.

M. Codescu et al. (Eds.). WADT 2014, LNCS 9463, pp. 90–109, 2015.
DOI: 10.1007/978-3-319-28114-8_6

The paper is structured as follows: Sect. 2 introduces examples from different fields. In Sect. 3 we briefly summarise some relevant notions from institution theory. The first approach to derived signature morphisms is to consider them to be ordinary signature morphisms into a definitional extension (Sect. 4). The second approach is to consider derived signature morphisms to be abstract substitutions that induce mappings on syntactic and semantic level (Sect. 5). The third approach is to consider institutional monads, which have derived signature morphisms as signature morphisms in their Kleisli institution (Sect. 5). We finish by discussing pros and cons and collecting open questions.

2 Examples

In specification theory derived signature morphisms may map between equivalent representations:

Example 1 (Boolean rings and algebras). It is well known, that Boolean rings and algebras are essentially the same thing. However, a mapping between these specifications has to cope with the fact, that the algebraic $\vee$ is an inclusive disjunction while the ring addition is an exclusive disjunction:

```
interpretation i : BooleanAlgebra to BooleanRing =
    ∧ ↦ λx,y.x·y
    ∨ ↦ λx,y.x+y+x·y
    ¬ ↦ λx.1+x
end

interpretation j : BooleanRing to BooleanAlgebra =
    · ↦ λx,y.x∧y
    + ↦ λx,y.(x∨y)∧¬(x∧y)
end
```

Note that operation symbols are mapped to λ-terms. The λ-variables open a context of variables for the subsequent terms. The number of λ-variables (or, for sorted logics, their sort string) must correspond to the arity of the operation symbol. further mote that the order λ-variables: $\lambda x, y.x$ is different from $\lambda x, y.y$.

Derived signature morphisms also play an important role in model-driven engineering (MDE). A problem that appears in practice when combining multiple models is that different models specify the same information differently.

Example 2. A related field of application is databases. Suppose we have two databases that we intend to use to store information about people, which were designed independently. We now wish to merge the information in the databases, We begin by merging the schema used to define the databases. To define the merge, we have to identify what the relationships are between the relation names and attributes of the two schema. Let the signature of DB_1 be $\langle\{Persons\}, \{Name, Gender, Age\}\rangle$, where *Persons* is the name of the database relation (table) intended to contain the information and *Name*, *Gender* and *Age* are attributes (columns) of this relation. Ditto DB_2, with signature $\langle\{\mathit{Male Female}\}$, $\{\mathit{Name}, \mathit{Bdate}, \mathit{Bplace}\}\rangle$, where we have two relations, *Male* and *Female*, and they both have attributes *Name*, $B(\mathit{irth})\mathit{date}$ and $B(\mathit{irth})\mathit{place}$. In order to create a suitable merge, we have to decide what matches what in the two schemas. Clearly the attribute *Gender* of DB_1 and the relations *Male* and *Female* are related, but cannot directly be matched as there is a type mismatch. If in DB_1 we define two new (derived) relations *Male* and *Female*, i.e., create a view of DB_1, both with attributes *Name* and *Age*, we will have "solved" the type mismatch problem for this aspect of the merge. These two relations can be defined as an extension of the original DB_1 schema by using an appropriate query in, say SQL. This is analogous to creating a definitional extension in FOL (see Example 5). So now we have DB_1' with the three relation names and the same attribute names.

Similarly, we can extend DB_2 with an extra attribute *Age* derived from *Bdate* via an appropriate query and obtain DB_2' with relations Male' and Female', and the extra attribute name *Age*. Now we define a span between DB_1' and DB_2', on the basis of which we can create the appropriate merged database schema by computing the colimit of the span. The database scheme DB at the apex of the span has signature $\langle \mathit{Male}, \mathit{Female}, \mathit{Name}, \mathit{Age}\rangle$. The maps connecting this scheme to DB_1' connect *Male* with *Male*, *Female* with *Female*, *Name* with *Name* and *Age* with *Age*. This is a Kleisli map between DB and DB_1, mapping DB to a definitional extension of DB_1. The maps connecting DB with DB_2' connect *Male* with Male' of DB_2', *Female* with Female', *Name* with *Name* and *Age* with *Age*. Again, this defines a Kleisli between DB and DB_2. Then, the corresponding pushout will be the "correct" merge of the two database schemas, avoiding redundancy in the merge and minimising the redundancy of data in the merged scheme.

Of course, having obtained the merged scheme, we would now want to merge the corresponding data. Database schemes correspond to theories and database instances correspond to models of theories. This framework could be used to derive the datamerge from the schema merge via amalgamation results. An alternative approach using a fibrational approach is outlined in [9].

Another field of interest are analogies. An analogy identifies common structures in the same or two different domains (source and target). In other words, an analogy basically consists of a common structural core that is instantiated in both, source and target.

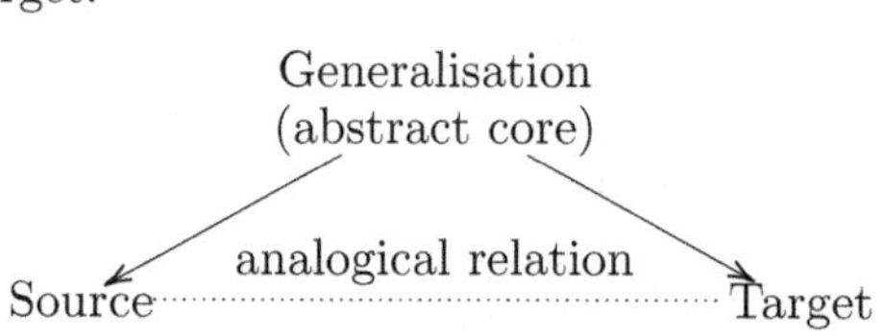

Hence, in logic an analogy can be formalised by giving a set of generalised formulas together with a pair of mappings, that map these formulas into the source and target respectively.[2] However, in many cases plain signature morphisms do not suffice to describe such a mapping, so derived signature morphisms offer a natural solution:

Example 3 (heat flow).[3] The heat flow analogy is well-known from physics education. The analogy is intended to introduce the concepts of heat and heat flow by comparing them to water and water flow. A simple description of this analogy may consist of the following observations: on the source side there are to vessels, a beaker and a vial, connected via a pipe. If the height of the water in the beaker is greater than the height of the water in the vial, water will flow and the height of the water in the beaker will decrease while the height of the water in the vial will increase. On the target side a metal bar is put into a cup of hot coffee. An ice cube is attached to the upper end of the bar. It is observed, that the coffee cools down while the ice heats up and finally melts. A logic-based representation of this description may contain the following formulas:

$$\begin{aligned}
&(\mathrm{G1})\ \mathit{connected}(A, B, C)\\
&(\mathrm{G2})\ \forall t_1 : \mathit{time},\, t_2 : \mathit{time} : t_2 > t_1\\
&\qquad \wedge T(A, t_1) > T(B, t_1)\\
&\qquad \rightarrow T(A, t_2) < T(A, t_1) \wedge T(B, t_2) > T(B, t_1)
\end{aligned}$$

$$\begin{aligned}
&(\mathrm{S1})\ \mathit{connected}(\mathit{beaker}, \mathit{vial}, \mathit{pipe})\\
&(\mathrm{S2})\ \forall t_1 : \mathit{time},\, t_2 : \mathit{time} : t_2 > t_1\\
&\qquad \wedge\ \mathit{height}(\mathit{in}(\mathit{water}, \mathit{beaker}), t_1)\\
&\qquad\quad > \mathit{height}(\mathit{in}(\mathit{water}, \mathit{vial}), t_1)\\
&\qquad \rightarrow \mathit{height}(\mathit{in}(\mathit{water}, \mathit{beaker}), t_2)\\
&\qquad\quad < \mathit{height}(\mathit{in}(\mathit{water}, \mathit{beaker}), t_1)\\
&\qquad \wedge\ \mathit{height}(\mathit{in}(\mathit{water}, \mathit{vial}), t_2)\\
&\qquad\quad > \mathit{height}(\mathit{in}(\mathit{water}, \mathit{vial}), t_1)
\end{aligned}$$

$$\begin{aligned}
&(\mathrm{T1})\ \mathit{connected}(\mathit{in}(\mathit{coffee}, \mathit{cup}), \mathit{ice_cube}, \mathit{bar})\\
&(\mathrm{T2})\ \forall t_1, t_2 : \mathit{time} : t_2 > t_1\\
&\qquad \wedge \mathit{temp}(\mathit{in}(\mathit{coffee}, \mathit{cup}), t_1)\\
&\qquad\quad > \mathit{temp}(\mathit{ice_cube}, t_1)\\
&\qquad \rightarrow \mathit{temp}(\mathit{in}(\mathit{coffee}, \mathit{cup}), t_2)\\
&\qquad\quad < \mathit{temp}(\mathit{in}(\mathit{coffee}, \mathit{cup}), t_1)\\
&\qquad \wedge\ \mathit{temp}(\mathit{ice_cube}, t_2)\\
&\qquad\quad > \mathit{temp}(\mathit{ice_cube}, t_1)
\end{aligned}$$

Here it is essential, that an object on the target side is matched to a vessel on the source side (but not to the water in the vessel). Hence, the following (derived) signature morphisms should be applied:

$$\begin{aligned}
\mathit{beaker} &\hookleftarrow A \mapsto \mathit{in}(\mathit{coffee}, \mathit{cup})\\
\mathit{vial} &\hookleftarrow B \mapsto \mathit{ice_cube}\\
\mathit{pipe} &\hookleftarrow C \mapsto \mathit{bar}\\
\lambda x \lambda h.\mathit{height}(\mathit{in}(\mathit{water}, x), t) &\hookleftarrow T \mapsto \lambda x \lambda h.\mathit{temp}(x, t)
\end{aligned}$$

3 Institutions

The study of derived signature morphisms can be carried out largely independently of the nature of the underlying logical system. We use the notion of

[2] Such an approach is used by the HDTP framework, described in [23].

[3] Simplified version from [23].

institution introduced by Goguen and Burstall [13] in the late 1970s (see [6] for a recent overview). It approaches the notion of logical system from a relativistic view: rather than treating the concept of logic as eternal and given, it accepts the need for a large variety of different logical systems, and instead asks about common principles shared across logical systems. A crucial feature of institutions is that logical structure is indexed by signature, and change of signature is accounted for by signature morphisms; this is of course what we need as a prerequisite for the concept of derived signature morphism.

Definition 1. *An* institution $\mathcal{I} = (\mathbb{S}ign, \mathbf{Sen}, \mathbf{Mod}, \models)$ *consists of*

- *a category* $\mathbb{S}ign$ *of signatures and signature morphisms,*
- *a functor* $\mathbf{Sen}\colon \mathbf{Sign} \to \mathbf{Set}$,[4] *giving a set* $\mathbf{Sen}(\Sigma)$ *of* Σ-sentences *for each signature* $\Sigma \in |\mathbf{Sign}|$, *and a function* $\mathbf{Sen}(\sigma)\colon \mathbf{Sen}(\Sigma) \to \mathbf{Sen}(\Sigma')$, *denoted by* $\sigma(\_)$, *that yields* σ-translation *of* Σ*-sentences to* Σ'*-sentences for each signature morphism* $\sigma\colon \Sigma \to \Sigma'$;
- *a functor* $\mathbf{Mod}\colon \mathbf{Sign}^{op} \to \mathbb{CAT}$,[5] *giving a category* $\mathbf{Mod}(\Sigma)$ *of* Σ-models *for each* $\Sigma \in |\mathbf{Sign}|$, *and a functor* $\mathbf{Mod}(\sigma)\colon \mathbf{Mod}(\Sigma') \to \mathbf{Mod}(\Sigma)$, *denoted by* $\_|_\sigma$, *that yields* σ-reducts *of* Σ'*-models for each signature morphism* $\sigma\colon \Sigma \to \Sigma'$; *and*
- *for each* $\Sigma \in |\mathbf{Sign}|$, *a* satisfaction relation $\models_{\mathcal{I},\Sigma} \subseteq |\mathbf{Mod}(\Sigma)| \times \mathbf{Sen}(\Sigma)$

such that for any signature morphism $\sigma\colon \Sigma \to \Sigma'$, Σ*-sentence* $\varphi \in \mathbf{Sen}(\Sigma)$ *and* Σ'*-model* $M' \in |\mathbf{Mod}(\Sigma')|$*:*

$$M' \models_{\mathcal{I},\Sigma'} \sigma(\varphi) \iff M'|_\sigma \models_{\mathcal{I},\Sigma} \varphi \qquad \textit{[Satisfaction condition]}$$

The satisfaction condition expresses that truth is invariant under change of notation and context.

Example 4. The institution *Prop* of propositional logic. Signatures are sets (of propositional variables), signature morphisms are functions. Models are valuations of propositional variables into $\{T, F\}$, model reduct is just composition of the given model with the corresponding signature morphism. Sentences are formed inductively from propositional variables by the usual logical connectives. Sentence translation means replacement of propositional variables along the signature morphism. Satisfaction is the usual satisfaction of a propositional sentence under a valuation. □

Example 5. The institution $FOL^=$ of many-sorted first-order logic with equality. Signatures are many-sorted first-order signatures, consisting of a set of sort and sorted operation and predicate symbols. Signature morphisms map sorts, operation and predicate symbols in a compatible way. Models are many-sorted first-order structures. Sentences are first-order formulas. Sentence translation means replacement of symbols along the signature morphism. A model reduct interprets a symbol by first translating it along the signature morphism and then

[4] The category **Set** has all sets as objects and all functions as morphisms.

[5] $\mathbb{CAT}$ is the quasi-category of all categories, where "quasi" means that it lives in a higher set-theoretic universe.

interpreting it in the model to be reduced. Satisfaction is the usual satisfaction of a first-order sentence in a first-order structure. □

Example 6. In [17], we have sketched an institution of database schemas. We here follow a naive approach: A database schema is essentially a FOL theory where (some of) the relation symbols correspond to the database relations and some of the sorts correspond to the database attributes. There may be axioms of the theory defining concepts like keys and so on. A morphism is simply a normal theory interpretation. Database instances are then just models over a fixed universe, with different universes defining different families of instances. An interesting point to note is that the usual relation algebra operators, like join, can be seen as patterns for defining endofunctors (in the category of FOL theories and morphisms) of the theory (database extension) that create definitional extensions of the database schema to which they are applied. As queries are compositions of such operators, a query is also an endofunctor defining a definitional extension of the scheme. □

Semantic entailment in an institution is defined as usual: for $\Gamma \subseteq \mathbf{Sen}(\Sigma)$ and $\varphi \in \mathbf{Sen}(\Sigma)$, we write $\Gamma \models \varphi$, if all models satisfying all sentences in Γ also satisfy φ.

An alternative definition of institution uses so-called 'rooms' (in the terminology of [12]), which capture the Tarskian notion of satisfaction of a sentence in a model:

Definition 2. *A* room $\mathcal{R} = (S, \mathcal{M}, \models)$ *consists of*

- *a set of S of* sentences*,*
- *a category $\mathcal{M}$ of* models*, and*
- *a binary relation $\models \subseteq |\mathcal{M}| \times S$, called the* satisfaction relation*.*

Then, morphisms between rooms are of course called corridors [12]:

Definition 3. *A* corridor $(\alpha, \beta)\colon (S_1, \mathcal{M}_1, \models_1) \to (S_2, \mathcal{M}_2, \models_2)$ *consists of*

- *a sentence translation function $\alpha\colon S_1 \to S_2$, and*
- *a model reduction functor $\beta\colon \mathcal{M}_2 \to \mathcal{M}_1$, such that*

$$M_2 \models_2 \alpha(\varphi_1) \textit{ if and only if } \beta(M_2) \models_1 \varphi_1$$

*holds for each $M_2 \in |\mathcal{M}_2|$ and each $\varphi_1 \in S_1$ (*satisfaction condition*).*

Since corridors compose and there are obvious identity corridors, rooms and corridors form a category $\mathbb{R}oom$. Then, an *institution* is just a functor $\mathcal{I}\colon \mathbb{S}ign \to \mathbb{R}oom$.

Relationships between institutions (and entailment systems) are captured mathematically by 'institution morphisms', of which there are several variants, each yielding a category under a canonical composition. For the purposes of this paper, institution morphisms [14] seem technically most convenient. For the notion of institutional monad introduced below, we also need 2-cells between institution morphisms, called modifications.

We use the representation of institutions as functors introduced above.

Definition 4. *Given institutions $I_1\colon \mathbb{S}ign_1 \to \mathbb{R}oom$ and $I_2\colon \mathbb{S}ign_2 \to \mathbb{R}oom$, an* institution morphism $(\Phi, \rho)\colon I_1 \to I_2$ *consists of a functor $\Phi\colon \mathbb{S}ign_1 \to \mathbb{S}ign_2$ and a natural transformation $\rho\colon I_2 \circ \Phi \to I_1$.*

Given institution morphisms $(\Phi, \rho)\colon I_1 \to I_2$ and $(\Phi', \rho')\colon I_1 \to I_2$, an institution morphism modification $\theta\colon (\Phi, \rho) \to (\Phi', \rho')$ *is just a natural transformation $\theta\colon \Phi \to \Phi'$ such that $\rho = \rho' \circ (I_2 \cdot \theta)$.*[6]

This leads to a 2-category $\mathbb{I}ns$ of institutions, morphisms and modifications.

Example 7. There is an institution morphism $\mu_1 : FOL^{=} \to Prop$. From a first-order signature, it only keeps the nullary predicates, which become propositional variables. Also from a first-order model, only the interpretations of the nullary predicates are kept. Moreover, there is an obvious inclusion of *Prop*-sentences into $FOL^{=}$-sentences. The satisfaction condition is easily shown. □

Example 8. Another institution morphism $\mu_2 : FOL^{=} \to Prop$ keeps *all* predicates from a first-order signature as propositional variables. From a first-order model, extract a valuation by mapping a predicate to true iff it is universally true. A propositional variable is translated to a sentence stating that the corresponding predicate holds universally. Again, the satisfaction condition is easily shown. □

Example 9. The inclusions $\iota_\Sigma : (\mu_1)_\Sigma \to (\mu_2)_\Sigma$ form a modification $\iota : \mu_1 \to \mu_2$. □

4 Derived Signature Morphisms Through Definitional Extensions

In this section, we develop a very general approach to derived signature morphisms, based on two assumptions: (1) signatures are replaced by *theories*, and (2) models can be *amalgamated.* While these assumptions and the idea of letting theory morphisms be targeted in some definitional extension is folklore, surprisingly little is known about the properties of this construction.

We start with colimits, which can be seen as a tool for combining and interconnecting systems, and amalgamation, which ensures that models can be combined along colimits. Amalgamation ensures further nice logical properties, e.g. laws for modularity [7], availability of institution-independent proof calculi for structured specifications [2,19] or well-behaved semantics for architectural specifications [22].

Definition 5. *A cocone for a diagram in $\mathbb{S}ign$ is* (weakly) amalgamable *if it is mapped to a (weak) limit in $\mathbb{CAT}$ under* Mod. *$\mathcal{I}$ (or* Mod*) admits* (finite) (weak) amalgamation *if (finite) colimits exists in $\mathbb{S}ign$ and colimiting cocones are (weakly) amalgamable, i.e. if* Mod *maps (finite) colimits to (weak) limits.*

[6] The original notion from [4] is a lax variant of this: a morphism $\rho \to \rho' \circ (I_2 \cdot \theta)$ is given instead of equality.

An important special case is pushouts: $\mathcal{I}$ *(or* Mod*) has* (weak) model amalgamation for pushouts, *if pushouts exist in* $\mathbb{S}ign$ *and are (weakly) amalgamable. More specifically, the latter means that for any pushout*

$$\begin{array}{ccc} \Sigma & \longrightarrow & \Sigma_1 \\ \downarrow & & \downarrow \\ \Sigma_2 & \longrightarrow & \Sigma_R \end{array}$$

in $\mathbb{S}ign$ *and any pair* $(M_1, M_2) \in \mathsf{Mod}(\Sigma_1) \times \mathsf{Mod}(\Sigma_2)$ *that is* compatible *in the sense that* M_1 *and* M_2 *reduce to the same* Σ*-model can be* amalgamated *to a unique (or weakly amalgamated to a not necessarily unique)* Σ_R*-model* M *(i.e., there exists a (unique)* $M \in \mathsf{Mod}(\Sigma_R)$ *that reduces to* M_1 *and* M_2*, respectively), and similarly for model morphisms.*

This specific explanation in terms of compatible families of models that can be amalgamated also generalises to arbitrary colimits.

For example, it is well-known [21] that

Proposition 1. *Both propositional logic and many-sorted first-order logic both have model amalgamation.*

In the sequel, we work in an arbitrary but fixed institution $\mathcal{I} = (\mathbb{S}ign, \mathbf{Sen}, \mathbf{Mod}, \models)$.

Definition 6. *A* theory *is a pair* $T = (\Sigma, \Gamma)$ *where* Γ *is a set of* Σ*-sentences. A* theory morphism $(\Sigma, \Gamma) \to (\Sigma', \Gamma')$ *is a signature morphism* $\sigma : \Sigma \to \Sigma'$ *such that* $\Gamma' \models_{\Sigma'} \sigma(\Gamma)$*. Let* $\mathbb{T}h(\mathcal{I})$ *denote this category. Each theory* (Σ, Γ) *inherits sentences from* $\mathsf{Sen}^{\mathcal{I}}(\Sigma)$*, while the models are restricted to those models in* $\mathsf{Mod}^{\mathcal{I}}(\Sigma)$ *that satisfy all sentences in* Γ*. It is easy to see that* $\mathcal{I}$ *maps theory morphisms to corridors in this way. By taking* $\mathbb{T}h(\mathcal{I})$ *as "signature" category, we arrive at the institution* $\mathcal{I}^{Th}$ *of theories.*

Definition 7. *A theory morphism* $\sigma \colon T_1 \to T_2$ *is* conservative, *if each* T_1*-model has a* σ*-expansion to a* T_2*-model; it is* definitional, *if each model has a unique such expansion. Definitional theory morphisms are also called* definitional extensions *and are denoted as* $T_1 \bullet\!\!\xrightarrow{\sigma} T_2$.

Definition 8. *A derived theory morphism* $(\sigma, \theta) \colon T_1 \to T_2$ *is given by an ordinary theory morphism* $\sigma \colon T_1 \to T_2'$ *into a definitional extension* $\theta \colon T_2 \bullet\!\!\longrightarrow T_2'$ *of* T_2.

Every theory morphism $\sigma \colon T_1 \to T_2$ is a derived theory morphism with respect to the identity $id \colon T_2 \bullet\!\!\longrightarrow T_2$. We can define reducts for arbitrary derived theory morphisms by first taking the unique θ-expansion and then taking σ-reduct.

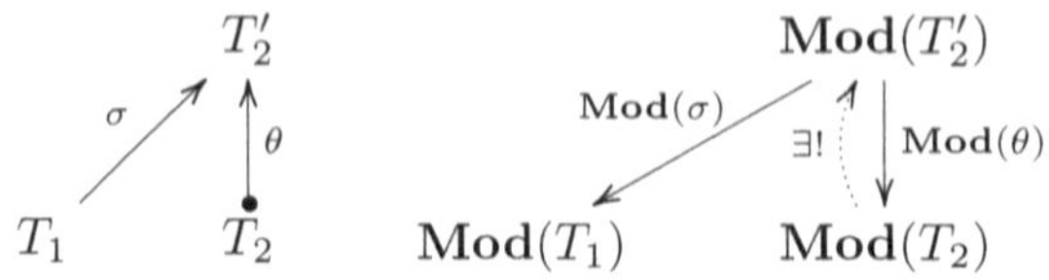

If we had based derived theory morphisms on conservative instead of definitional extensions, reducts would exist but generally would not be unique due to the possibility to have several different θ-expansions.

Example 10. In the setting of Example 1, we can construct a derived theory morphism $T_{\mathrm{ring}} \to T_{\mathrm{algebra}}$ via definitional extension T'_{algebra}:

$$\Sigma'_{\mathrm{algebra}} := \Sigma_{\mathrm{algebra}} \cup \{+\}$$
$$\Gamma'_{\mathrm{algebra}} := \Gamma_{\mathrm{algebra}} \cup \{x + y = (x \vee y) \wedge \neg(x \wedge y)\}$$

One can then define an ordinary signature morphism $\sigma : \Sigma_{\mathrm{ring}} \to \Sigma'_{\mathrm{algebra}}$ by mapping $\cdot \mapsto \wedge$ and $+ \mapsto +$.

Note that there is a caveat: adding defined symbols generally can change the notion of model morphism. For example, consider a $FOL^=$-signature with a binary predicate symbol Q. Then adding a unary predicate symbol P with definition

$$P(x) \Leftrightarrow \forall y.Q(x, y)$$

is indeed a definitional extension. However, not every model morphism for Q will also preserve P. As a consequence, derived theory morphisms in general do not provide reducts for model morphisms. Hence, in the sequel, we have to make the following

General Assumption. Model morphisms are compatible with definitional extensions, which means that given a definitional extension $\sigma : T_1 \to T_2$, every T_1-model morphism $h_1 : M_1 \to M_1'$ has a unique expansion to a T_2-model morphism $h_2 : M_2 \to M_2'$, where M_2 is the unique expansion of M_1 and M_2' that of M_1'.

One simple way to achieve this property is to dispense with non-trivial model morphisms and use discrete model categories. Another way is to restrict formulas in derived theory morphisms to those that are compatible with all model morphisms. Both ways have certain drawbacks, but there is no easy solution to this problem.

Further note that in general a derived theory morphism $T_1 \to T_2$ does not provide a translation of T_1-sentences to T_2-sentences. This means that we will arrive at a category of theories and derived theory morphisms which form a *specification frame*, which is given by a category **Spec** of (abstract) specifications (or theories), with semantics given by a model functor $\mathbf{Mod} \colon \mathbf{Spec}^{op} \to \mathbb{CAT}$. The terminology follows [3], the concept appeared earlier as "specification logic" in [10,11]). As before, functions $\mathbf{Mod}(\sigma)$, for $\sigma : T_1 \to T_2$ in **Spec**, will be called reducts and denoted by $\_|_\sigma$.

Our derived theory morphisms are similar to morphisms in the category $Cospan(\mathcal{I}^{Th})$. However, the latter category has severe drawbacks: generally, only some colimits exist, see [1]. Moreover, the equivalence used for cospans, isomorphism of intermediate objects, is much too fine-grained for our purposes. In general, there are many choices for a derived signature morphism due to underdetermination of the intermediate theory T_2': arbitrary symbols may be added to the signature Σ_2' and equivalent formulations of the sentences in Γ_2' can be chosen. Such modifications do not change the essence of the morphism, i.e. induced mappings on model level. The following definition will account for this fact:

Definition 9. *Two derived theory morphisms* $(\sigma_1, \theta_1), (\sigma_2, \theta_2) \colon T_1 \to T_2$ *are* equivalent, *if their induced model reduct maps are equal.*

Proposition 2. *In am institution $\mathcal{I}$ with model amalgamation for pushouts, derived theory morphisms compose. This leads to a category $\mathbb{D}er(\mathcal{I})$ of derived theory morphisms up to equivalence.*

Proof. The composition $(\sigma_2, \theta_2) \circ (\sigma_1, \theta_1)$ is given by $(\sigma \circ \sigma_1, \theta \circ \theta_2)$, where the rhombus is a pushout:

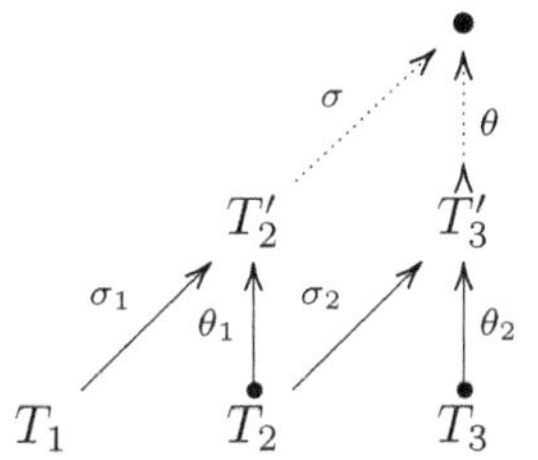

In order to show that $\theta \circ \theta_2$ is definitional, it suffices to show that θ is (θ_2 is by definition). Let M_3' be T_3'-model. Since θ_1 is definitional, $M_3'|_{\sigma_2}$ has a unique expansion to a T_2'-model M_2'. Then the unique amalgamation of M_3' and M_2' gives a the unique desired expansion of M_3'. This shows the folklore fact that definitional extensions are preserved by pushouts.

Composition is well-defined, because different pushouts always lead to equivalent derived theory morphisms. □

Note that in case that $\theta_1 = id$, the composition simplifies to $(\sigma_2 \circ \sigma_1, \theta_2)$. Further note that for the verification of commutativity of diagrams in $\mathbb{D}er(\mathcal{I})$, it is often easier to compose model reduct maps; this avoids the computation of the above pushout.

Altogether, we arrive at

Theorem 1. *For a given institution $\mathcal{I}$, theories and their models together with derived theory morphisms and their reducts from a specification frame $\mathcal{I}^{Der}$.*

A central result is to establish the existence of colimits in $\mathbb{D}er(\mathcal{I})$:

Theorem 2. *In an institution with model amalgamation, the category $\mathbb{D}er(\mathcal{I})$ of derived theory morphisms is cocomplete.*

Proof. First note that colimits lift from signatures to theories [13], so we can assume that the category of theories $\mathbb{T}h(\mathcal{I})$ is cocomplete.

The initial theory 0 is also initial in $\mathbb{D}er(\mathcal{I})$: Given any theory T, the derived theory morphism from 0 to T is $(!_T, id_T)$. Concerning its uniqueness, note that by model amalgamation, $\mathsf{Mod}(0)$ is a singleton, which means that there all derived theory morphisms starting from 0 are equivalent.

Concerning non-empty products, given a set of theories $(T_i)_{i\in I}$, its coproduct $\coprod_I T_i$ in the category of theories lifts to $\mathbb{D}er(\mathcal{I})$. The coproduct injections in $\mathbb{D}er(\mathcal{I})$ are $T_i \xrightarrow{\amalg_i} \coprod_I T_i \xleftarrow{id} \coprod_I T_i$. To show the universal property, let $T_i \xrightarrow{\tau_i} U_i \xleftarrow{\theta_i} \bullet T$ be a cocone in $\mathbb{D}er(\mathcal{I})$. Let $(C, (\mu_i \colon U_i \to C)_{i\in I})$ be the colimit of $(U \bullet\!\xrightarrow[\theta_i]{} U_i)_{i\in I}$ in $\mathbb{T}h(\mathcal{I})$. Then let $\tau \colon \coprod_I T_i \to C$ be $[\mu_i]_I \circ \coprod_I \tau_i$. Pick some $i_0 \in I$. The mediating morphism from the colimit to the cocone is then $(\tau, \mu_{i_0} \circ \theta_{i_0}) \colon \coprod_I T_i \to T$ ($\mu_{i_0} \circ \theta_{i_0}$ is equal to $\mu_i \circ \theta_i$ for any $i \in I$). $\mu_{i_0} \circ \theta_{i_0}$ is definitional: any T-model M has unique θ_i-expansions $M_i \in \mathsf{Mod}(U_i)$ for $i \in I$. Then M together with the M_i form a compatible family of models for $(U \bullet\!\xrightarrow[\theta_i]{} U_i)_{i\in I}$. By model amalgamation, this family has a unique amalgamation to a model M_C of the colimit C such that $M_C|_{\mu_i} = M_i$. Now $\mu_{i_0} \circ \theta_{i_0}$ commutes with the cocones, because (1) given a T-model, its unique C-expansion reduces via μ_i to its unique U_i-expansion and (2) the left triangle commutes by definition of τ.

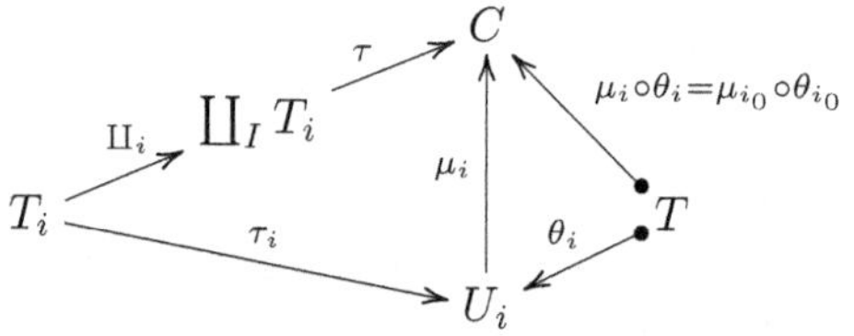

To show its uniqueness, assume that there is another morphism $(\lambda, \theta) \colon \coprod_I T_i \to T$ with $(\lambda, \theta) \circ (\amalg_i, id) = (\tau_i, \theta_i)$, which means that the two model reduct maps from T to T_i are the same:

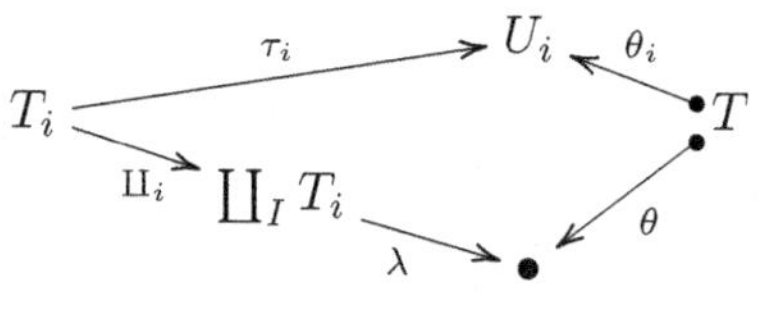

But then $(\lambda, \theta) = (\tau, \mu_{i_0} \circ \theta_{i_0})$ by model amalgamation:

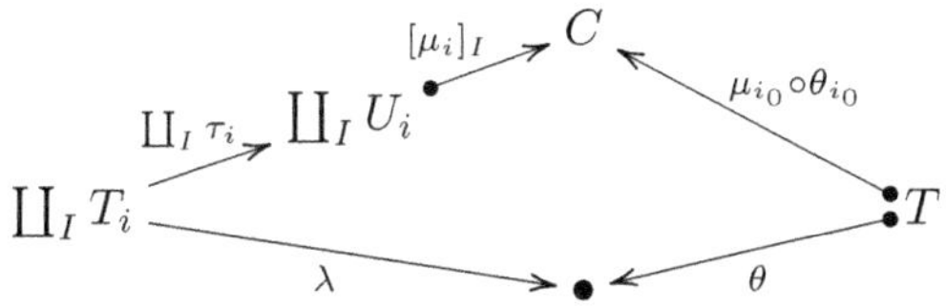

It remains to treat coequalisers. Given a pair of parallel derived theory morphisms $(\sigma_1, \theta_1), (\sigma_2, \theta_2) \colon T \to U$, its coequaliser in $\mathbb{D}er(\mathcal{I})$ is given by (μ, id_C), which is obtained as the following colimit of theories:

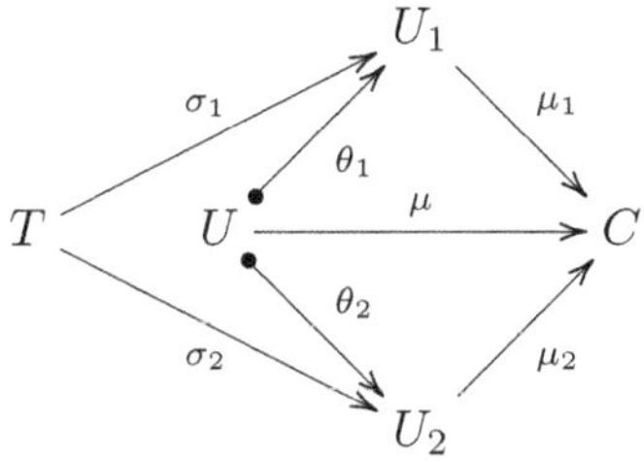

(*)

By definitionality of θ_1 and θ_2 and model amalgamation, (μ, id_C) is an epi in $\mathbb{D}er(\mathcal{I})$. Concerning the universal property, consider any cocone in $\mathbb{D}er(\mathcal{I})$ $(\tau, \theta) \colon U \to D$. Take the pushout of theories shown in the left square

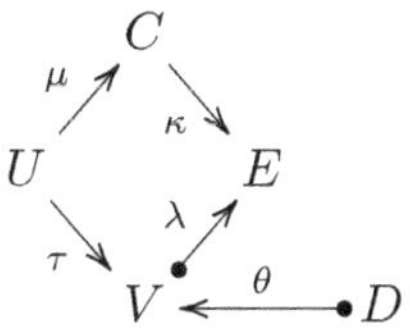

Then $(\kappa, \lambda \circ \theta) \colon C \to D$ is the mediating morphism. To establish definitionality of λ, consider any V-model M. Let $M_i \in \mathsf{Mod}(U_i)$ by the unique θ_i-expansion of $M_U := M|_\tau$ $(i = 1, 2)$. Since (τ, θ) is a cocone, $(\tau, \theta) \circ (\sigma_1, \theta_1) = (\tau, \theta) \circ (\sigma_1, \theta_1)$. Now consider the D-model $M_D := M|_\theta$. Then

$$M_1|_{\sigma_1} = M_U|_{(\sigma_1, \theta_1)} = M_D|_{(\tau,\theta)\circ(\sigma_1,\theta_1)} = M_D|_{(\tau,\theta)\circ(\sigma_2,\theta_2)} = M_U|_{(\sigma_2,\theta_2)} = M_2|_{\sigma_2}$$

Let us denote this model by M_T. Thus, (M_T, M_U, M_1, M_2) is a compatible family of models for the diagram $(*)$ above (without the C), which by model amalgamation can be amalgamated to a C-model M_C. Let M_E be the amalgamation of M_C and M. Then M_E is the needed λ-expansion of M. Its uniqueness follows from those of the amalgamations.

From the diagram above we easily get that $(\kappa, \lambda \circ \theta) \circ (\mu, id_C) = (\tau, \theta)$. Uniqueness follows since (μ, id_C) is an epi in $\mathbb{D}er(\mathcal{I})$. □

A natural follow-up question is whether the specification frame $\mathcal{I}^{Der}$ admits amalgamation. Under some mild assumption, the answer is positive:

Theorem 3. *$\mathcal{I}^{Der}$ admits amalgamation whenever $\mathcal{I}$ does.*

Proof. Since $\mathbb{D}er(\mathcal{I})$ inherits coproducts from $\mathcal{I}$, also amalgamation lifts. Concerning coequalisers, in the notion of diagram $(*)$ above, let M_U a U-model such that $M_U|_{(\sigma_1,\theta_1)} = M_D|_{(\tau,\theta)\circ(\sigma_1,\theta_1)}$. Let $M_i \in \mathsf{Mod}(U_i)$ be the unique θ_i-expansion of M_U $(i = 1, 2)$. Then $(M_U|_{\theta_1}, M_1, M_2, M_U)$ is a compatible family for $(*)$ without C. This family can be uniquely amalgamated to a model M_C of C, which is the desired amalgamation in $\mathbb{D}er(\mathcal{I})$. Uniqueness follows from that of the amalgamation in I and that of definitional extensions. □

We have defined equivalence of derived theory morphisms using a semantic condition that is undecidable in general. However, for specific institutions, one can do better. For example, in $FOL^=$, one can restrict definitional extensions to those that are given by explicit definitions of predicate and function symbols (i.e. by equivalence to a formula or equality to a term). Then one can use syntactic equality of symbol definitions in order to decide equivalence of derived signature morphisms. This yields an efficiently decidable approximation of semantic equivalence.

A more syntactic notion of equivalence of derived theory morphisms would require definitional extensions to be monic. Then, two derived theory morphisms $(\sigma_1, \theta_1)\colon T_1 \to T_2$ and $(\sigma_2, \theta_2)\colon T_1 \to T_2$ are said to be equivalent, if there is a theory T' and commutative diagrams as follows:

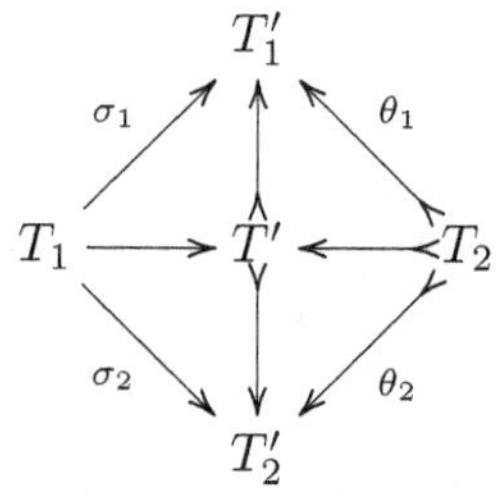

Under suitable assumptions, coproducts exist; however, coequalisers do not. This is why we have chosen the semantic notion of equivalence above.

5 Derived Signature Morphisms as Abstract Substitutions

The second approach to derived signature morphisms is to consider them to be a special kind of abstract substitution in the sense of [5,6]. Such Σ-substitutions generalise the idea of substitutions found in many logics to arbitrary institutions. The extension of a signature by a set of variables is expressed by a signature morphism $\chi : \Sigma \to \Sigma'$, leading to the following definition (we give a version that makes use of rooms and corridors):

Definition 10. *For any signature Σ of an institution I, with two "extensions" $\chi_1\colon \Sigma \to \Sigma_1$ and $\chi_2\colon \Sigma \to \Sigma_2$, a* Σ-substitution *$\chi_1 \to \chi_2$ is a corridor $\rho\colon I(\Sigma_1) \to I(\Sigma_2)$ that preserves Σ, i.e. the following diagram commutes*

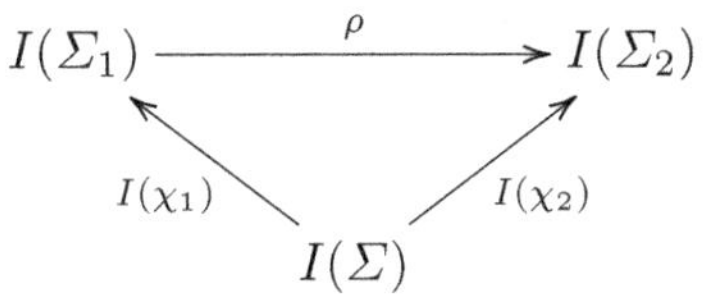

The idea of this definition is the existence of sentence translations and model reducts between extensions of a signature. This makes it a very general concept, that covers besides classical first-order substitution also second-order substitutions in $FOL^=$ and also derived signature morphisms. [6, 99f] demonstrates this for the case of $FOL^=$:[7] for a given base signature Σ, a derived signature $\Phi(\Sigma)$ is constructed. There exists a canonical embedding $\eta : \Sigma \to \Phi(\Sigma)$. It is then shown that sentences over the derived signature can be translated to sentences over the base signature and that a model for the base signatures provide a model for the derived signature, i.e. there is a corridor ρ from $I(\Phi(\Sigma))$ to $I(\Sigma)$. In summary, such a derivation is a Σ-substitution from η to id (this could also be expressed simply by saying that ρ is a retraction of $I(\eta)$ in the category $\mathbb{R}oom$).

Given such a derivation (Φ, η, ρ) for a signature Σ_2, a derived signature morphism from Σ_1 to Σ_2 is defined as an ordinary signature morphism $\sigma\colon \Sigma_1 \to \Phi(\Sigma_2)$. The substitution condition assures, that sentence translation and model reduction hold for the underlying base category, i.e. Σ_1-sentences can be translated to Σ_2-sentences, and Σ_2-models can be reduced to Σ_1-models along σ by detour over $\Phi(\Sigma_2)$:

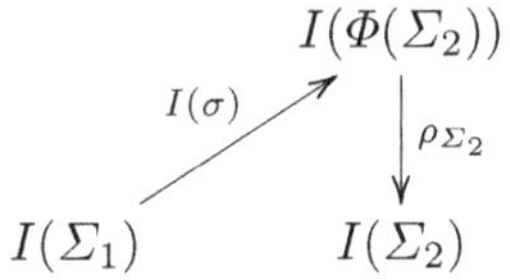

Analysing the above example, one can find the following ingredients that seem to be essential to introduce the concept of derivation in an institution I:

- a general way to construct *derived signatures*, i.e. a functor $\Phi : \mathbb{S}ign \to \mathbb{S}ign$
- a *canonical embedding* from the base signature to its derivation, i.e. a natural transformation $\eta\colon id \to \Phi$
- a *translation* (corridor) from the "derived logic" to "simple logic", i.e. a natural transformation $\rho : I \circ \Phi \to I$
- compatibility of the embedding and the translation, expressed by the condition $\rho \circ (I \cdot \iota) = id$

Using the language of institution morphisms and institution morphism modifications, this amounts to saying:

Definition 11. *A* derivation *for an institution I consists of*

- *an institution morphism $T = (\Phi^T, \rho^T)\colon \mathcal{I} \to \mathcal{I}$ and*
- *an institution morphism modification $\eta\colon id \to T$*

[7] We give only a brief summary here, simplifying and adapting notation.

This allows to introduce derived signature morphisms between arbitrary signatures of $\mathbb{S}ign$. However, a shortcoming of this approach is that derived signature morphisms can not be composed in an obvious way. This will be addressed in the next section.

Before closing this section, it should be remarked that one can use derivations as an alternative way to introduce substitutions at an abstract level: reconsidering the above situation, where a signature Σ and two extensions $\chi_1 : \Sigma \to \Sigma_1$ and $\chi_2 : \Sigma \to \Sigma_2$ are given, a derivation-based *Σ-substitution* $\chi_1 \to \chi_2$ is defined as a derived signature morphism $\sigma : \Sigma_1 \to \Sigma_2$ that preserves Σ, i.e. it makes following diagram of signature morphisms commute:

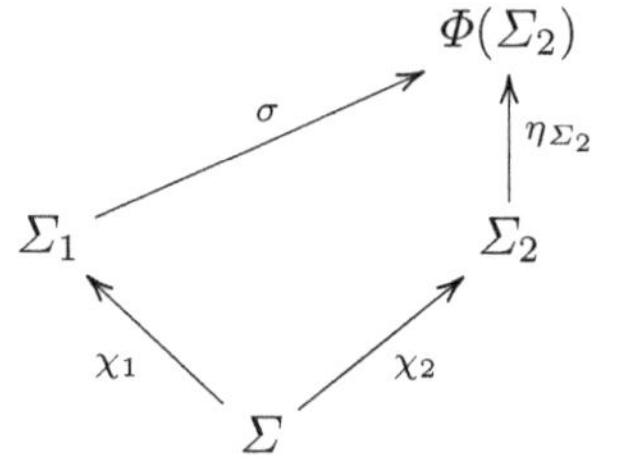

For example, in the case of $FOL^=$, consider extensions χ_1 and χ_2 of a signature Σ by first-order variables, i.e. new 0-ary operator names. Then a substitution μ replaces variables of Σ_1 by terms over Σ_2, i.e. symbols from a derivation $\Phi(\Sigma_2)$. Higher-order substitutions can be obtained by extending the signature with higher-order variables, i.e. new operators names with arity > 0. This notion of derivation-based Σ-substitution, can also be used to introduce an abstract notion of unification: given a set of Σ_1-sentences S, a unifier is a derived signature morphism $\mu : \Sigma_1 \to \Sigma_1$ (that preserves Σ) such that the induced mapping on sentence level, i.e. $\rho_{\Sigma_1} \circ \mathsf{Sen}^I(\mu) : \mathsf{Sen}^I(\Sigma_1) \to \mathsf{Sen}^I(\Sigma_1)$ maps S on a singleton set.

This notion of substitution is more specific than the above one, since every derivation-based Σ-substitution σ is a Σ-substitution in the sense of Definition 10: the induced corridor $\rho = \rho_{\Sigma_2} \circ I(\sigma)$ obviously preserves $I(\Sigma)$. The advantage of the derivation-based approach is, that it anchors the corridor ρ in a mapping on signature level in a natural way, while it stills seems general enough to cover most interesting cases.

6 Derived Signature Morphisms Through Kleisli Institutions

The approaches of the previous sections have some drawbacks: In the definitional extension approach, sentences cannot be translated along derived theory morphisms, while substitution-based derived signature morphisms do not compose. In this section, we remedy these problems by introducing for each signature Σ, a *signature of terms* $\Phi^T(\Sigma)$, where T is a suitable *monad*. Then, a derived signature morphism $\sigma\colon \Sigma_1 \to \Sigma_2$ is an ordinary signature morphism $\sigma\colon \Sigma_1 \to \Phi^T(\Sigma_2)$.

The monad needs to interact with the structure of the institution. This leads to the notion of *institutional monad.*

Definition 12. *An* institutional monad $\mathcal{T} = (T, \eta, \mu)$ *is a monad in* $\mathbb{I}ns$ *(see [18] for the notion of monad in a 2-category), which amounts to*

- *an institution* $\mathcal{I}$,
- *an institution morphism* $T = (\Phi^T, \rho^T)\colon \mathcal{I} \to \mathcal{I}$,
- *an institution morphism modification* $\eta\colon id \to T$, *and*
- *an institution morphism modification* $\mu\colon T \times T \to T$,

such that the usual laws of a monad are satisfied:

$$\begin{array}{ccc} T & \xrightarrow{\eta_T} & T \times T \\ {\scriptstyle T\eta}\downarrow & \searrow\!\!\searrow & \downarrow{\scriptstyle \mu} \\ T \times T & \xrightarrow{\mu} & T \end{array} \qquad\qquad \begin{array}{ccc} T \times T \times T & \xrightarrow{\mu_T} & T \times T \\ {\scriptstyle T\mu}\downarrow & & \downarrow{\scriptstyle \mu} \\ T \times T & \xrightarrow{\mu} & T \end{array}$$

By selecting the signature component only, an institutional monad $\mathcal{T}$ gives rise to an ordinary a monad, which we denote by $\mathcal{T}^{\mathbb{S}ign}$.

Example 11 (monad over the first-order logic institution). Let T be the institution morphism $(\Phi^T, \rho^T)\colon FOL^= \to FOL^=$ with $\rho^T_\Sigma = (\alpha^T_\Sigma, \beta^T_\Sigma)$.

- $\Phi^T(\Sigma)$ adds terms $\lambda x_1 : s_1, \ldots x_n : s_n.t$ as n-ary operations and terms $\lambda x_1 : s_1, \ldots x_n : s_n.\varphi$ (where φ is a formula) as n-ary predicates;
- $\alpha^T_\Sigma\colon Sen(\Phi^T(\Sigma)) \to Sen(\Sigma)$ β-reduces all application of λ-term operations and predicates;
- $\beta^T_\Sigma\colon Mod(\Sigma) \to Mod(\Phi^T(\Sigma))$ interprets λ-term operations and predicates in $\beta^T_\Sigma(M)$ as $(a_1, \ldots, a_n) \mapsto M(t)[x_i \mapsto a_i]$
- $\eta_\Sigma\colon \Sigma \to \Phi^T(\Sigma)$ is the obvious inclusion;
- $\mu_\Sigma\colon \Phi^T(\Phi^T(\Sigma)) \to \Phi^T(\Sigma)$ collapses two levels of λ-terms into one.

The notion of Kleisli category for a monad can be generalised to institutions in following way:

Definition 13. *Given an institutional monad* $\mathcal{T} = (T\colon \mathcal{I} \to \mathcal{I}, \eta, \mu)$, *its* Kleisli institution $\mathcal{I}_\mathcal{T}$ *is the Kleisli object of* $\mathcal{T}$ *in* $\mathbb{I}ns$, *which amounts to*

- *the signature category of* $\mathcal{I}_\mathcal{T}$ *is the Kleisli category of the monad* $\mathcal{T}^{\mathbb{S}ign}$,
- *given a signature* $\Sigma \in \mathcal{T}^{\mathbb{S}ign}$, $\mathcal{I}_\mathcal{T}(\Sigma)$ *is just* $\mathcal{I}(\Sigma)$, *and*
- *given a signature morphism* $\sigma\colon \Sigma_1 \to \Sigma_2 \in \mathcal{T}^{\mathbb{S}ign}$ *(which is a signature morphism* $\sigma\colon \Sigma_1 \to \Phi^T(\Sigma_2)$ *in* $\mathcal{I}$*),* $\mathcal{I}_\mathcal{T}(\sigma)$ *is given by*

$$\mathcal{I}(\Sigma_1) \xrightarrow{\mathcal{I}(\sigma)} \mathcal{I}(\Phi^T(\Sigma_2)) \xrightarrow{\rho^T_{\Sigma_2}} \mathcal{I}(\Sigma_2)$$

providing sentence translation and model reduct for Kleisli morphisms.

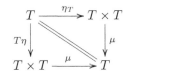

The institution independent notions of logical consequence, theory etc. and corresponding results of course also apply to the Kleisli institution; in particular, Kleisli theory morphisms preserve logical consequence.

Contrary to the statement in [9], colimits are not necessarily lifted from the base signature category to the Kleisli signature category:

Proposition 3. *If the base institution $\mathcal{I}$ has signature coproducts, then does the Kleisli institution $\mathcal{I}_\mathcal{T}$. However, coequalisers (and therefore also e.g. pushouts) are generally not lifted to the Kleisli institution.*

Proof. If $\Sigma_i \xrightarrow{\amalg_i} \coprod_I \Sigma_i$ is a coproduct in the signature category of $\mathcal{I}$, then $\Sigma_i \xrightarrow{\amalg_i} \coprod_I \Sigma_i \xrightarrow{\eta_{\coprod_I \Sigma_i}} T(\coprod_I \Sigma_i)$ is a coproduct in the signature category. See also (2.1) in [24].

Concerning coequalisers, consider the category of derived signature morphisms of standard first-order logic. Take a parallel pair of arrows where a binary function symbol is mapped a) to $\lambda x, y : s.x$ and b) to $\lambda x, y : s \mapsto y$. Then there is no coequalisers, since it would have to equate $x, y \mapsto x$ with $x, y \mapsto y$. (A span with no pushout can be obtained in a similar way.) □

Note that the negative situation in Proposition 3 can be remedied in some cases. For the example given in the proof, a coequaliser exists in the category of *derived theory morphisms up to equivalence.* In this pushout, an axiom $\forall x, y : s.x = y$ is added. This category can be defined as follows:

Definition 14. *For the institution of many-sorted first-order logic, the category of* derived theory morphisms up to equivalence *has theories as objects. Morphisms are derived signature morphisms that map axioms to theorems, taken up to an equivalence. Two derived theory morphisms are equivalent iff they map a given symbol to terms that are provably equal.*

Proposition 4. *The category of derived* theory *morphisms up to equivalence for $FOL^=$ has colimits.*

Proof. For coproducts, use Proposition 3. The coequaliser of a pair

$$U \underset{\sigma_2}{\overset{\sigma_1}{\rightrightarrows}} V$$

is obtained in the base signature category by

$$V \xrightarrow{q} Q \xrightarrow{\eta_Q} TW$$

where Q is the quotient of V by the congruence

$$\sigma_1(s) \equiv \sigma_2(s) \ (s \in sorts(U)).$$

Then on sorts, $q \circ \sigma_1 = q \circ \sigma_2 =: q'$. Moreover, W is Q augmented by axioms

$$\forall x_1 : q'(s_1), \ldots x_n : q'(s_n).\alpha(\sigma_1(f)(x_1, \ldots, x_n) = \sigma_2(f)(x_1, \ldots, x_n))$$

for each operation symbol $f : s_1 \ldots s_n \to s$ in U and axioms

$$\forall x_1 : q'(s_1), \ldots x_n : q'(s_n).\alpha(\sigma_1(p)(x_1, \ldots, x_n) \Leftrightarrow \sigma_2(p)(x_1, \ldots, x_n))$$

for each predicate symbol $p : s_1 \ldots s_n$ in U. Recall from Example 11 that the effect of α is that all applications of $\sigma_1(f)$ (resp. $\sigma_1(p)$) to terms are β-reduced.

Now the Kleisli morphism $\eta_Q \circ q \colon V \to W$ equalises σ_1 and σ_2: for sorts, this is done by q, and for operation and predicate symbols, this follows from the axioms in W (noting that provably equal symbols are identified). Given any Kleisli morphism $h \colon V \to X$ equalising σ_1 and σ_2, define $k \colon W \to X$ by $k(q(s)) = h(s)$ on sorts, and $k(f) = h(f)$ for operation and predicate symbols. In both cases, well-definedness follows from $h \circ \sigma_1 = h \circ \sigma_2$. □

7 Conclusions

We have introduced several approaches to derived signature (resp. theory) morphisms. The first approach, using definitional extensions, is very general and works in any institution with model amalgamation for pushouts. While models can be reduced against derived signature morphisms, the drawback is that sentences cannot be translated along them. The second approach remedies this problem axiomatically: model reducts and sentence translation are required to exist. Moreover, powerful Herbrand theorems relate queries and substitutions [5,6]. The third approach is more specific about the nature of derived signature morphisms: they are obtained through a Kleisli construction in an institutional monad, which provides a more precise (abstract) description of what derived signature morphisms are.

Generally, it turns out that coproducts lift easily to the derived case, while coequalizers are more difficult. The problem is that derived signature morphisms are too powerful to admit coequalisers directly, because in a coequalisers, they can be used to equate arbitrarily complex terms. The trick to still obtain coequalisers is to pass from signature to theory morphisms and impose some suitable quotient on the latter. For the approach of definitional extensions, we can obtain coequalizers by working with theory morphisms and consider derived theory morphisms up to semantic equivalence, while a stronger (more syntactic) equivalence does not work. For the particular Kleisli institutions of the natural institutional monad for many-sorted first-order logic, we can obtain coequalizers by adding suitable equations. It is an interesting open question whether and how this can be generalised to an arbitrary institution.

There are still open questions concerning the relationship between the notion introduced via definitional extensions and the one using the Kleisli construction. One can ask, if (and under which conditions) it is possible to define a "definitional extension institutional monad", in which a derivation consists of the colimit of all "suitable" definitional extensions. It seems promising to consider syntactic definitional extensions, i.e. those that induce a mapping on the sentence level that is compatible with the model expansion. Another interesting point concerns the development of a general way to construct institutional monads, that would provide a kind of canoncial derivation. Here the idea of a charter [12] may provide a starting point.

On a more general level, this approach shows again, that notions from basic category theory (monads and Kleisli construction) can be adopted to institutions and lead to useful concepts there. It naturally leads to the question, if related notions, like the Eilenberg-Moore construction, can give raise to meaningful applications in an institutional setting as well.

References

1. nLab: Span. http://ncatlab.org/nlab/show/span
2. Borzyszkowski, T.: Logical systems for structured specifications. Theor. Comput. Sci. **286**, 197–245 (2002)
3. Cornelius, F., Baldamus, M., Ehrig, H., Orejas, F.: Abstract and behaviour module specifications. Math. Struct. Comput. Sci. **9**(1), 21–62 (1999)
4. Diaconescu, R.: Grothendieck institutions. Appl. Categorical Struct. **10**, 383–402 (2002)
5. Diaconescu, R.: Herbrand theorems in arbitrary institutions. Inf. Process. Lett. **90**, 29–37 (2004)
6. Diaconescu, R.: Institution-Independent Model Theory. Birkhäuser, Basel (2008)
7. Diaconescu, R., Goguen, J., Stefaneas, P.: Logical support for modularisation. In: Huet, G., Plotkin, G. (eds.) Proceedings of a Workshop on Logical Frameworks (1991)
8. Diskin, Z., Kadish, B.: A graphical yet formalized framework for specifying view systems. In: Manthey, R., Wolfengagen, V. (eds.) Advances in Databases and Information Systems 1997, Proceedings of the First East-European Symposium on Advances in Databases and Information Systems, ADBIS 1997, St Petersburg, 2–5 September 1997 (1997)
9. Diskin, Z., Maibaum, T., Czarnecki, K.: Intermodeling, queries, and kleisli categories. In: de Lara, J., Zisman, A. (eds.) Fundamental Approaches to Software Engineering. LNCS, vol. 7212, pp. 163–177. Springer, Heidelberg (2012)
10. Ehrig, H., Baldamus, M., Cornelius, F., Orejas, F.: Theory of algebraic module specification including behavioral semantics and constraints. In: Nivat, M., Rattray, C., Rus, T., Scollo, G. (eds.) AMAST 1991. Workshops in Computing, pp. 145–172. Springer, Heidelberg (1992)
11. Ehrig, H., Baldamus, M., Orejas, F.: New concepts of amalgamation and extension for a general theory of specifications. In: Bidoit, M., Choppy, C. (eds.) Abstract Data Types 1991 and COMPASS 1991. LNCS, vol. 655. Springer, Heidelberg (1993)
12. Goguen, J.A., Burstall, R.M.: A study in the foundations of programming methodology: specifications, institutions, charters and parchments. In: Poigné, A., Pitt, D.H., Rydeheard, D.E., Abramsky, S. (eds.) Category Theory and Computer Programming. LNCS, vol. 240, pp. 313–333. Springer, Heidelberg (1986)
13. Goguen, J.A., Burstall, R.M.: Institutions: Abstract model theory for specification and programming. J. Assoc. Comput. Mach. **39**, 95–146 (1992). Predecessor in: LNCS, vol. 164, pp. 221–256 (1984)
14. Goguen, J.A., Roşu, G.: Institution morphisms. Formal Aspects Comput. **13**, 274–307 (2002)
15. Goguen, J.A., Thatcher, J.W., Wagner, E.G.: An initial algebra approach to the specification, correctness and implementation of abstract data types. In: Yeh, R.T. (ed.) Current Trends in Programming Methodology - vol. IV: Data Structuring, pp. 80–149. Prentice-Hall (1978)

16. Honsell, F., Longley, J., Sannella, D., Tarlecki, A.: Constructive data refinement in typed lambda calculus. In: Tiuryn, J. (ed.) FOSSACS 2000. LNCS, vol. 1784, pp. 161–176. Springer, Heidelberg (2000)
17. Kutz, O., Mossakowski, T., Lücke, D.: Carnap, goguen, and the hyperontologies: logical pluralism and heterogeneous structuring in ontology design. Log. Univers. **4**(2), 255–333 (2010)
18. Lack, S.: A 2-categories companion. In: Baez, J.C., May, J.P. (eds.) Towards Higher Categories. The IMA Volumes in Mathematics and its Applications, vol. 152, pp. 105–191. Springer, New York (2010)
19. Mossakowski, T., Autexier, S., Hutter, D.: Development graphs - proof management for structured specifications. J. Logic Algebraic Program. **67**(1–2), 114–145 (2006). http://www.sciencedirect.com/science?_ob=GatewayURL&_origin=CONTENTS&_method=citationSearch&_piikey=S1567832605000810&_version=1&md5=7c18897e9ffad42e0649c6b41203f41e
20. Sannella, D.T., Burstall, R.M.: Structured theories in LCF. In: Protasi, M., Ausiello, G. (eds.) CAAP 1983. LNCS, vol. 159, pp. 377–391. Springer, Heidelberg (1983)
21. Sannella, D., Tarlecki, A.: Foundations of Algebraic Specification and Formal Software Development. Monographs in Theoretical Computer Science. Springer, Berlin (2012)
22. Schröder, L., Mossakowski, T., Tarlecki, A., Klin, B., Hoffman, P.: Amalgamation in the semantics of casl. Theor. Comput. Sci. **331**(1), 215–247 (2005)
23. Schwering, A., Krumnack, U., Kühnberger, K.U., Gust, H.: Syntactic principles of heuristic-driven theory projection. J. Cogn. Syst. Res. **10**(3), 251–269 (2009). Special Issue on Analogies - Integrating Cognitive Abilities
24. Szigeti, J.: On limits and colimits in the Kleisli category. Cahiers de Topologie et Géométrie Différentielle Catégoriques **24**(4), 381–391 (1983)

Use Case Analysis Based on Formal Methods: An Empirical Study

Marcos Oliveira Jr.(✉), Leila Ribeiro, Érika Cota, Lucio Mauro Duarte, Ingrid Nunes, and Filipe Reis

PPGC – Institute of Informatics, Federal University of Rio Grande do Sul (UFRGS), PO Box 15.064, Porto Alegre, RS 91.501-970, Brazil
{marcos.oliveira,leila,erika,lmduarte, ingridnunes,freis}@inf.ufrgs.br

Abstract. *Use Cases (UC)* are a popular way of describing system behavior and represent important artifacts for system design, analysis, and evolution. Hence, UC quality impacts the overall system quality and defect rates. However, they are presented in natural language, which is usually the cause of issues related to imprecision, ambiguity, and incompleteness. We present the results of an empirical study on the formalization of UCs as Graph Transformation models (GTs) with the goal of running tool-supported analyses on them and revealing possible errors (treated as open issues). We describe initial steps for a translation from a UC to a GT, how to use an existing tool to analyze the produced GT, and present some diagnostic feedback based on the results of these analyses and the possible level of severity of the detected problems. To evaluate the effectiveness of the translation and of the analyses in identifying problems in UCs, we applied our approach on a set of real UC descriptions obtained from a software developer company and measured the results using a well-known metric. The final results demonstrate that this approach can reveal real problems that could otherwise go undetected and, thus, help improve the quality of the UCs.

Keywords: Use cases · Graph transformation · Empirical study · Model analysis

1 Introduction

Use Cases (UC) [3] are a popular model for documenting software expected behavior. They are used in different software processes, not only for requirement documentation and validation, but also as specifications for system design, verification, and evolution. Hence, they are important reference points within the software development process. In current practice, UC descriptions are typically informally documented using, in most cases, natural language in a predefined structure. Being informal descriptions, UCs might be ambiguous and imprecise.

This work is partially supported by the VeriTeS project (FAPERGS and CNPq).

M. Codescu et al. (Eds.): WADT 2014, LNCS 9463, pp. 110–130, 2015.
DOI: 10.1007/978-3-319-28114-8_7

This may result in a number of specification problems that can be propagated to later development phases and jeopardize the overall system quality [1]. In fact, it is well-known that most software faults are introduced during the specification phase [12]. Nevertheless, it is important to keep UC descriptions in a format familiar to the stakeholders, since they must be involved in the UC definition. Thus, the verification of UCs normally corresponds to manual inspections and walkthroughs [11]. Because the analysis is manual, detecting incompleteness and recognizing ambiguities is not a trivial task. Since software quality is highly dependent on the quality of the specification, cost-effective strategies to decrease the number of errors in UCs are crucial.

Strategies for the formalization of UCs have already been proposed, such as [7,8,10,15]. Many of them assume a particular syntax for UC description tailored for their particular formalisms. This limits the expression of requirements in terms of the stakeholders language and, in some cases, also restrains the semantics of the UC. Moreover, whereas current design techniques are mostly data-driven, which delays control-flow decisions until later phases, many of the used formalisms model UCs as sequences of actions, which may neglect data-related issues. Our aim is to keep the expressiveness of a description in natural language and use a formalism for modeling/analysing UCs that is flexible enough to represent the semantics defined by stakeholders at a very abstract level. Moreover, we advocate that the translation from a UC to a formal model should be performed in a systematic way, guided by well-defined steps (possibly aided by tools), such that the model can be obtained without an expert in the formalism (because the expertise is embedded in the predefined translation process). This is fundamental for the adoption of formal methods in practice.

In this paper, we investigate the suitability of Graph Transformation (GT) [5,14] as a formal model to describe and analyze UCs. Some reasons for choosing GT are: the elements of a UC can be naturally represented as graphs; it is a visual language; the semantics is very simple yet expressive; GT is data-driven; there are various static and dynamic analysis techniques available for GT, as well as tools to support them. We work towards an approach that integrates UC formalization and tool-supported analysis, with the objective of improving the quality of UCs. As the formalization requires a precise description of the behavior described in the UC, the process of translating it into a formal model may already reveal errors. The goal is to define a sequence of steps to guide the process of building the formal model, executing analyses, and evaluating the results in terms of the level of severity of errors. Diagnostic feedback should also be provided, indicating possible actions to solve the detected problems through modification of the original UC. Hence, the process should, iteratively and gradually, improve an initial UC and generate, as result, not only a more precise UC, which can still be presented to non-technical stakeholders and be readily used without affecting the usual development process, but also a corresponding formal model that can be refined and used in subsequent design activities. This paper presents the first steps towards such a process, presenting an outline of the idea and an empirical evaluation of the effectiveness of the translation and of the analyses in identifying problems in UCs. We applied our approach on a set of real UC descriptions obtained from a software development company and measured the results using a well-known metric. The final results demonstrate that this approach can reveal real problems that could otherwise go undetected and, thus, help improve the quality of the UCs.

This paper is organized as follows: Sect. 2 presents the necessary background information and details of the translation from UCs to GTs, as well as a detailed description of each step of our approach applied to a running a example; Sect. 3 presents the settings of the conducted empirical study; Sect. 4 presents an analysis and discussion of results; Sect. 5 discusses threats to the validity of our work; Sect. 6 presents a comparative analysis of our technique in relation to some similar techniques; and Sect. 7 concludes the paper and discusses future work.

2 Modeling UCs Using GTs

2.1 Background

Use Cases a *Use Case (UC)* defines a contract between stakeholders of a system, describing part of the system behavior [3]. The main purpose of a UC description is the documentation of the expected system behavior and to ease the communication between stakeholders, often including non-technical people, about required system functionalities. For this reason, the most usual UC description is the textual form. A general format of a UC contains a unique name, a primary actor, a primary goal, and a set of sequential steps describing the successful interaction between the primary actor and the system towards the primary goal. A sequence of alternative steps are often included to represent exception flows. Pre- and post-conditions are also listed to indicate, respectively, conditions that must hold before and after the UC execution.

Figure 1 depicts an example of UC of a bank system in a typical textual format, describing the log in operation executed by a bank client. We explain our approach using this UC as example.

Graph Transformations. The formalism of *Graph Transformations (GT)* [5,14] is based on defining states of a system as graphs and state changes as rules that transform these graphs. Due to space limitations, in this section, we only provide an informal overview of the notions used in this paper. For formal definitions, see e.g. [14]. Examples of graphs, rules and their analysis are presented in the following subsections.

Graphs are structures that consist of a set of nodes and a set of edges. Each edge connects two nodes of the graph, one representing a source and another representing a target. A *total homomorphism* between graphs is a mapping of nodes and edges that is compatible with sources and targets of edges. Intuitively, a total homomorphism from a graph *G1* to a graph *G2* means that all items (nodes and edges) of *G1* can be found in *G2* (but distinct nodes/edges of *G1* are not necessarily distinct in *G2*). If we have a graph, say *TG*, that represents all possible (graphical) types that are needed to describe a system, a total homomorphism *h* from any graph *G* to *TG* would associate a (graphical) type to each

Use Case Specification (original)

Number	1	
Name	Log into *ATM*	
Summary	User logs into ATM	
Priority	5	
Preconditions	User has *bank card* and registered *password*	
Postconditions	User receives *menu* of available *ATM* operations	
Primary Actor(s)	Bank *Customer*	
Secondary Actor(s)	Customer *Accounts Database*	
Trigger	Only option on ATM	
Main Scenario	**Step**	**Action**
	1	*System* asks for a *Bank card*
	2	*User* inserts *card*
	3	*System* asks for *password*
	4	*User* enters *password*
	5	*System* validates user's *card* and *password* and display *menu* of operations
Extensions	**Step**	**Branching Action**
	5a	*System* notifies *user* that *password* is invalid
	5b	*System* exits option
Open Issues		

Fig. 1. Login Use Case description.

item of G. We call this triple $\langle G, h, TG\rangle$ a *typed graph*, and TG is called a *type graph* (that is, nodes of TG describe all possible types of nodes of a system, and edges of TG describe possible relationships between these types).

A *Graph Rule* describes a relationship between two graphs. It consists of: a *left-hand side (LHS)*, which describes items that must be present for this rule to be applied; a *right-hand side (RHS)*, describing items that will be present after the application of the rule; and a *mapping from LHS to RHS*, which describes items that will be preserved by the application of the rule. This mapping must be compatible with the structure of the graphs (i.e., a morphism between typed graphs) and may be partial. Items that are in the LHS and are not mapped to the RHS are *deleted*, whereas items that are in the RHS and are not in the image of the mapping from the LHS are *created*. We also assume that rules do not merge items, that is, they are injective.

A *GT System* consists of a type graph, specifying the (graphical) types of the system, and a set of rules over this type graph that define the system behavior. The application of a rule r to a graph G is possible if an image of the LHS of r is found in G (that is, there is a total typed-graph morphism from the LHS of

r to G). The result of a rule application deletes from G all items that are not mapped in r and adds the ones created by r.

Our analysis of GTs is based on concurrent rules and critical pairs, two methods of analysis independent from the initial state of the system and, thus, they are complementary to any other verification strategy based on initial states (such as testing), detailed further ahead.

2.2 UC Formalization and Verification Strategy

Figure 2 depicts the proposed UC formalization and verification strategy, which is divided into four main phases. Starting from a textual description of the UC, the first phase (*UC Data Extraction phase*) is to identify entities (Step 1) and actions (Step 2) that will be part of the formal model. Then, basic verifications can be performed regarding the consistency of the extracted information (*Primary Verifications phase*). We look for inconsistencies that might affect or even prevent the construction of the GT model such as entities or conditions that are mentioned but never used, actions or effects of an action that are not clearly defined, and so on. If inconsistencies are detected, the UC must be rewritten to eliminate them or the analyst can annotate the problem as an open issue to be resolved later on. When no basic inconsistencies are found, the GT can then be generated (*GT Generation phase*). In this process, conditions and effects of actions are modeled as states (graphs) in Step 3. Then, in Step 4, a type graph is built through the definition of a graphical representation of the artifacts generated in Steps 1 and 3. After that (Step 5), each UC step is modeled as a transition rule from one state (graph) to another, using the structures defined in Steps 3 and 4.

Having the GT, a series of automatic verifications (based on concurrent rules, conflict analysis, and dependency analysis) can be performed to detect possible problems (*UC Analysis phase*). We use the AGG tool [18] to perform the automatic analyses on the GT model. All detected issues are annotated as *open issues (OIs)* along with the solutions (when applicable). With this approach, any design decision made over an OI can be documented and tracked back to the original UC. Through analysis, it is possible to verify whether the pre- and post-conditions were correctly included in the model, whether there are conflicting and/or dependent rules, what is the semantics of a detected conflict or dependency, and whether these results were expected or not. One important point is that, during the process of representing the UC in the formal model, clarifications and decisions about the semantics of the textual description must be made. Annotated OIs force the stakeholders to be more precise and explicit about tacit knowledge and unexpressed assumptions about system invariants and expected behavior.

Open issues are classified according to their severity level: code Yellow (⚠) indicates a warning, meaning a minor problem that can probably be solved by a single person; code Orange ( ) indicates a problem that requires more attention and probably a definition/confirmation from the stakeholders; code Red ( ) indicates a serious issue that requires a modification in the UC description.

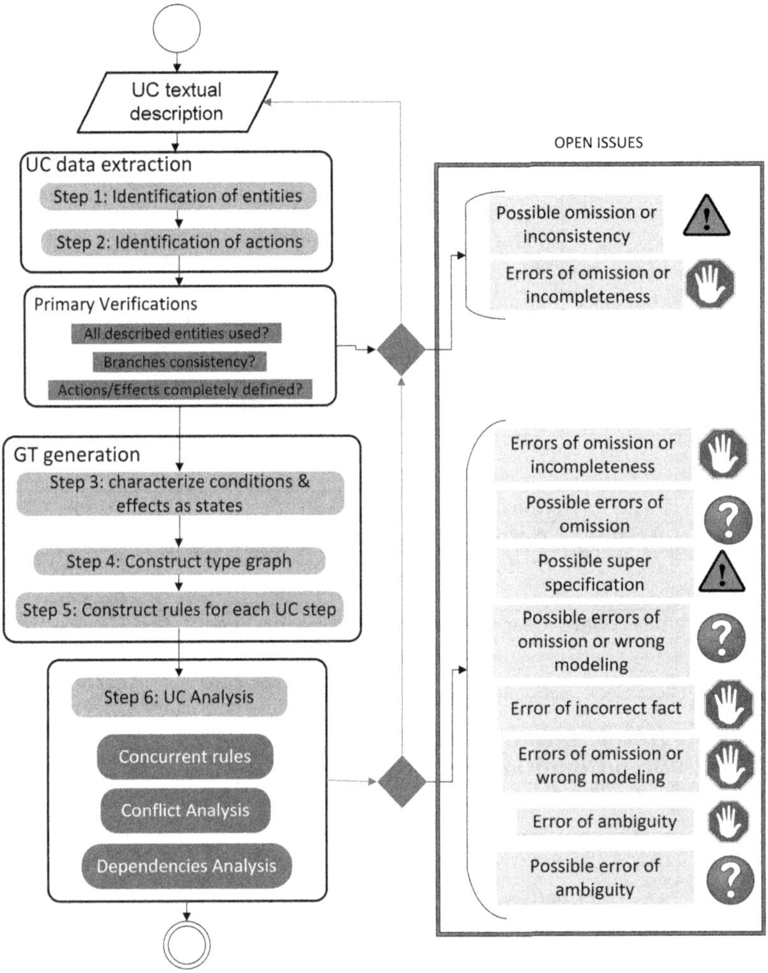

Fig. 2. Overview of the UC formalization and verification strategy.

UC 1	Step 1	Artifacts	*List of Entities: User, ATM, System, Bank card, Password*

Fig. 3. UC1 - Step 1

Below, we describe the steps of our UC formalization and verification approach and the possible OIs that can be derived from them, illustrating the results for the login UC (UC1 - Fig. 1).

Step 1 - *Identification of entities*: The analyst manually identifies in the UC text all the entities involved in UC. Figure 3 shows the result for the example.

Table 1. Primary verification steps

Open issue	Verification	Problem	Severity level	Possible action
OI.1	An entity listed in Step 1 is not used (as actor or involved) in any action	Different names for the same entity or entities used in pre-/post-conditions are not used in the steps of the UC	⚠ Yellow	Analyze whether this is actually what is intended,
OI.2	A branching condition is not used in any action	The description of the actions may be too abstract	⚠ Yellow	Analyze whether this is actually what is intended
OI.3	The effect of an action is not clearly defined	Ambiguous description or omission	✋ Red	Provide more details in the UC description

UC 1 Step 2

Artifacts

Table of Actions UC1

Action	Actor	Involved	Conditions	Effect
askCard	System	IO	—	Display msg asking card
insertCard	User	System, Card	1.System asks for card 2.User has card	Card becomes connected to system
...				

Table of Branch Conditions UC1

Step	Condition	Value: Step
5	User's card and password are validated	true: 5. false: 5a

Open Issues

⚠ *ATM never appeared in the actions table* (OI.1)

✋ *"exits option" - step 5b - not clear* (OI.3)

⚠ *Branching conditions was not used* : nothing was said about how validation should be carried out. (OI.2)

Fig. 4. UC1 - Step 2

Step 2 - Identification of actions: This step defines a *Table of Actions*, containing an entry for each action in the UC.

Actions that perform input/output operations involve a special entity called *IO*. Considering the possibility of alternative paths described in the UC, a *Table of Branch Conditions* is also defined.

Based on these two tables, three basic verifications can be performed and may raise open issues, as detailed in Table 1. Figure 4 shows the result of this step to the example UC. Only part of the Table of Actions is presented. As a result, three open issues were raised.

Step 3 - *Modeling conditions and effects as states:* In this step, it must be explicitly defined how to describe the conditions and effects listed in the *Table* *of Actions*, as well as the pre- and post-conditions of the UC in terms of nodes and edges of a graph. The resulting table is called *Table of Conditions/Effects*. At the same time, we build a *Table of Operations* that is used in these formal definitions, with two predefined operations *Input* and *Output*. The tables resulting from this step are illustrated in Fig. 5.

UC 1 Step 3 — Artifacts

(Part of) Table of Conditions Effects UC1

Action	**Condition/Effect**	**Characterization**
pre	User has bank card and registered password	
insertCard, askCard	System displays msg asking for card / System asks for card	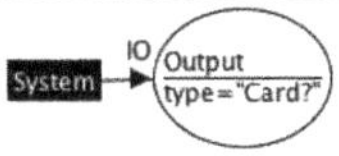
...		

Table of Operations UC1

OPN	**Src**	**Tgt**	**RetVal**	**Pars**	**UsedIn**
Output	System	—	—	type: String	askCard, askPwd, validate&display, validate&notiy
Input	—	System	—	type: String pwd: Password	enterPwd

Fig. 5. UC1 - Step 3

Step 4 - *Construction of the Type graph:* The nodes of the type graph are the entities (**Step 1**) and operations (**Step 3**). The arcs are the relationships that were necessary to characterize the conditions/effects. If attributes of nodes were used to characterise the conditions/effects, they must also be part of the type graph. Figure 6 shows the type graph for our running example.

UC 1 Step 4 — Artifacts

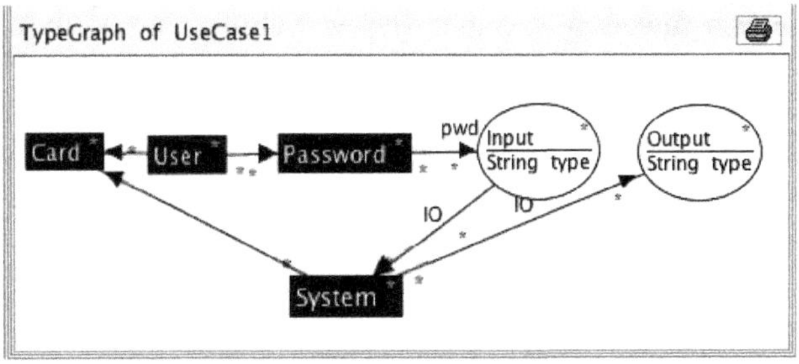

Fig. 6. UC1 - Step 4

Step 5 - *Construction of rules:* Rules that formally describe the behavior of the UC are constructed. For each action listed in the *Table of Actions*, we build

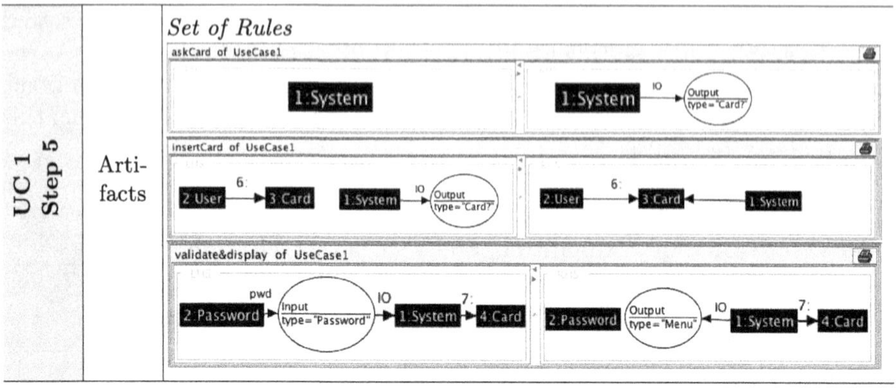

Fig. 7. UC1 - Step 5

a rule having as left-hand side (LHS) the graph that describes the conditions that must be true for this action to occur. The graphs corresponding to each condition are already described in the *Table of Conditions/Effects*, hence it is only necessary to merge them appropriately. Analogously, the right-hand side (RHS) of the rule is built using the effects of each action. Some rules of the example UC are shown in Fig. 7.

Step 6 - *Use case analysis:* The following analysis techniques may be performed in any order. The result from these analyses are usually complementary.

6.1. Use Case Effect: We first define *rule sequences (RSs)* that represent the execution of each possible path of the UC. RSs are just sequences of rules (defined in **Step 5**) that implement the execution of each scenario of the UC. Based on each RS, we build a single rule, called *concurrent rule*, which shows the effect of the whole UC in one step. This concurrent rule allows us to check whether the overall effect is really the desired one. Table 2 presents the analysis performed on the RSs and possible resulting open issues.

This analysis makes it explicit: (i) everything that is required for the UC to execute (LHS of the rule); and, (ii) the overall effect of the UC (RHS of the rule). To build the concurrent rule, the rules of the UC are joint by dependencies and, therefore, if some items are forgotten, this might lead to the impossibility of building the concurrent rule using all rules of the UC (and, thus, we might discover errors in the description of the UC steps as rules). Figure 8 shows the result of this step for our UC example.

6.2. Conflict Analysis (critical pairs): This type of analysis technique tells us which steps are mutually exclusive, that is, it pinpoints the choice points of the system. Table 3 presents the verifications based on critical pairs analysis and the possible resulting open issues. The result of the conflict critical-pair analysis is a *Conflict Matrix*, having rules as rows and columns, where each cell is filled with a number indicating how many items of a rule (row) are in conflict with items of another rule (column).

Table 2. Verification steps on rules sequences

Open issue	Verification	Problem	Severity level	Possible action
OI.4	A concurrent rule (for any alternative path in the UC) cannot be built using all the rules in the corresponding RSs	Items generated by some rule and used by another one may be missing by omission or modeling error	Red	Review the rules
OI.5	Multiple concurrent rules are built for a single UC scenario	Multiple instances of one or more entities are possible, leading to different (possibly unexpected) ways of combining the rules of the UC	Red	Check dependencies between rules to find unexpected subpaths in the UC behavior
OI.6	UC pre-conditions are not a subgraph of the LHSs of the concurrent rules	Pre-conditions may include unnecessary items	⚠ Yellow	Remove unused pre-conditions from the UC text
OI.7	The LHS of a concurrent rule is not a subgraph of the UC pre-conditions	UC requires something that is not explicitly stated in the pre-conditions	Orange	Identify the RS in problematic concurrent rule and check whether all actions in this path were correctly modeled. If model is correct, check for missing pre-conditions.
OI.8	Post-conditions of an alternative path of the UC are not contained in the RHS of the corresponding concurrent rule	Some rule is not generating a required item (by UC omission or modeling mistake)	Red	Check the rules. If all rules seem to be correct, post-conditions might be too strong.
OI.9	The RHS of a concurrent rule is not contained in the corresponding UC post-condition	Some rule is not deleting a required item (by UC omission or modeling mistake)	Red	If the rules seem to correctly describe each action, post-conditions might be too weak

A value Zero in a cell means there is no conflict between two rules. The conflicts are only between items of the LHSs of the rules. The results of this step for our example are presented in Fig. 9.

6.3. Dependency Analysis (critical pairs): Similarly to critical pair analysis, (potential) dependency analysis is independent of an initial situation and is performed by building a *Dependency Matrix*. It shows relationships between rules and can be used to check whether the dependencies that we intuitively expected to occur are actually there. The verifications based on this matrix are presented in Table 4.

UC 1 Step 6.1	Arti-facts	*Concurrent Rules* 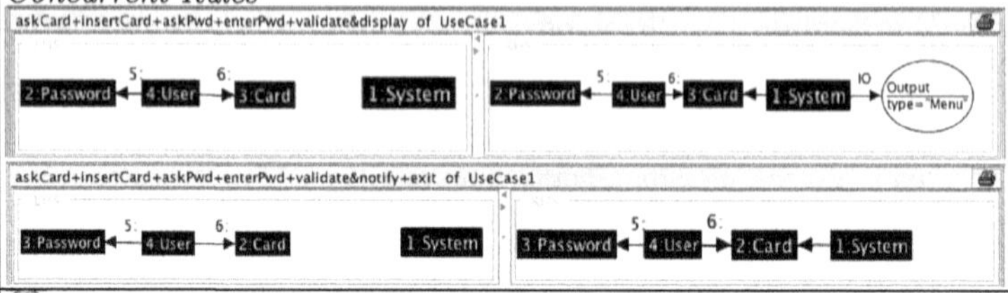
	Open Issues	*Missing preconditions*: System must exist (OI.7) *System did not release card in error execution*: review UC and rule (OI.9)

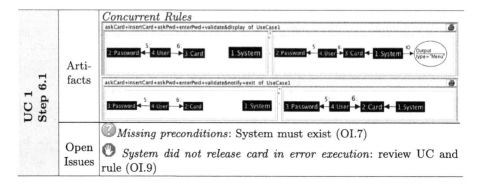

Fig. 8. UC1 - Step 6.1

Table 3. Verification steps based on critical pairs analysis

Open issue	Verification	Problem	Severity level	Possible action
OI.10	A rule is not conflicting with itself	The rule could be applied an arbitrary number of times	⚠ Yellow	Analyze whether this is actually the intended behavior
OI.11	There is no conflict between rules that represent the branching points of the UC behavior	Non-deterministic behavior: any alternative path can be taken no matter the condition	Red	Revise the conditions (LHSs) associated with rules representing alternative paths in the UC
OI.12	Conflicts between rules other than the ones described above (with itself and branch points)	These conflicts represent branches in system execution that must be explicitly stated in the UC (and in the model) as an alternative path	Orange	Revise the conflicting rules

Table 4. Verification steps based on dependencies analysis

Open issue	Verification	Problem	Severity level	Possible action
OI.13	Dependencies listed do not represent dependencies that are desired in the system	Possible omission in the UC description or a modeling error	⚠ Yellow	Check the RHS of a rule and the LHS of the other rule that depends on the first one
OI.14	An expected dependency between rules does not appear	Possible omission in the UC description or a modeling error.	⚠ Yellow	Check the rules involved

UC 1 Step 6.2	Arti-facts	*Conflicts Matrix*							
		Minimal Conflicts							
		Show							
		first \ second	1	2	3	4	5	6	7
		1 askCard	0	0	0	0	0	0	0
		2 insertCard		5	0	0	0	0	0
		3 askPwd	0	0	0	0	0	0	0
		▪ ▪							
	Open Issues	⚠ *No self-conflicts on rules askCard and askPwd*: Add a status attribute to prevent unexpected applications of these rules (OI.10)							

Fig. 9. UC1 - Step 6.2

UC 1 Step 6.3	Arti-facts	*Dependenciess Matrix*							
		Minimal Dependencies							
		Show							
		first \ second	1	2	3	4	5	6	7
		1 askCard	0	1	0	0	0	0	0
		▪ ▪ ▪							
		7 exit	0	0	0	0	0	0	0
	Open Issues	⚠ *askCard does not depend on exit*: Rule exit must re-establish the initial conditions (changing the status variable accordingly) (OI.14)							

Fig. 10. UC1 - Step 6.3

Figure 10 shows the result of this step for our example UC. If two rules that we would like to occur always in some specific order are shown to be independent, then they could actually occur in any order, which represents a possible problem.

As a final result, we obtain a UC textual description more accurate and complete, as shown in Fig. 11, after all open issues have been analysed.

Currently, most of the steps are manual and some of them are carried out aided by tools. In steps 1 and 2, the analysis is purely manual, however, in steps 3, 4 and 5, the analyst should use a tool such as AGG, which helps build a formal model of the system by supporting the visual construction of graph grammars. This tool is also very useful in step 6 to perform the analysis of critical pairs and the generation of concurrent rules, which are not trivial processes.

3   Study Settings

Considering the methodology presented in the previous section, we now detail the study conducted to evaluate its usefulness for UC formalization and its effectiveness for tool-supported analysis of UCs in order to detect real and potential problems. This empirical study was crucial to obtain concrete evidences that using GTs in the context of UCs promotes quality.

In order to adequately evaluate our approach, we followed the principles of experimental software engineering [19]. We first present our study goal in

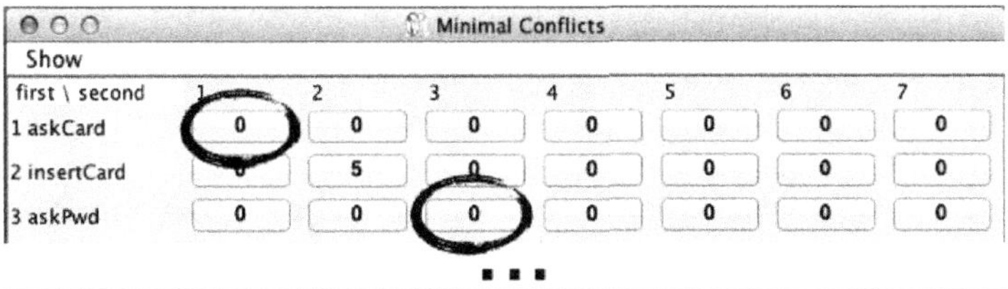

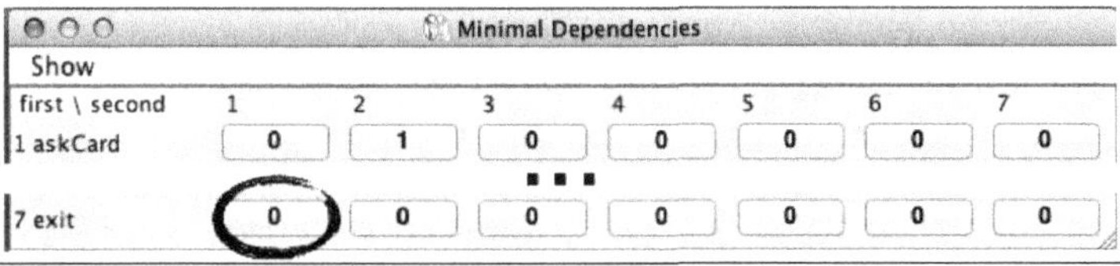

Use Case Specification (after verification)

Number	1
Name	Log into System via ATM
Summary	User logs into System via ATM
Priority	5
Preconditions	User has bank card and registered password and the system is running
Postconditions	User receives menu of available System operations
Primary Actor(s)	Bank Customer
Secondary Actor(s)	Customer Accounts Database

Trigger	Only option on ATM	
Main Scenario	**Step**	**Action**
	1	System asks for a Bank card
	2	User inserts card
	3	System asks for password
	4	User enters password
	5	System validates user's card and password and display menu of operations
Extensions	**Step**	**Branching Action**
	5a	System notifies user that password is invalid
	5b	System exits option and goes back to step 1. System releases the card.
Open Issues		

Fig. 11. Login Use Case description after verification.

Table 5. Goal definition.

Element	Our study goal
Motivation	To understand the usefulness of GTs to improve the quality of UCs
Purpose	Evaluate
Object	The effectiveness of using GTs to identify problems in UCs
Perspective	From a perspective of the researcher
Scope	In the context of a single real software development project

Table 5, which follows the GQM template [2]. Based on the definition of our goal, we derived two research questions, which we aim to answer with our study.

RQ-1. Are system analysts able to detect problems in their own UC descriptions without additional support?

RQ-2. How effective is our GT-based approach in identifying problems in UCs?

In order to answer these research questions, we followed the steps detailed in Sect. 3.1, which describes the study procedure. In Sect. 3.2, we introduce the software development project that is the target system of our study.

3.1 Procedure

1 - Analysis of UCs by System Analyst. In order to answer **RQ-1**, we requested a system analyst responsible for the creation of the UC descriptions, to carefully revise them, and point out problems, such as ambiguity, imprecision, omission, incompleteness, and inconsistency. This analyst has more than three years of experience in software projects with varying lengths (from a few weeks to years) with documentation describing the entire architecture of the solution, including artifacts such as class and sequence diagrams and use cases. If the system analyst found any problem, then such problems should ideally be identified by our approach. However, there was no guarantee the system analyst would be able to identify all existing problems. Therefore, the system analyst was a *"sound but not complete" oracle*, i.e. all problems identified were real problems but not necessarily all of the problems that existed in the UCs would be detected.

2 - UC Formalization. Given a set of 5 UCs, we performed the steps detailed in Sect. 2 to formalize them using GTs and used the AGG tool to analyze them. As a result, we detected some OIs.

3 - Evaluation of Detected Open Issues. After identifying open issues using our GT-based approach, we evaluated whether they were real problems in the analyzed UCs. If a detected OI had been pointed out as a real problem by the system analyst in the first step of our procedure, then it was definitely a real problem. Otherwise, the system analyst was requested to analyze the OI and verify whether it was an actual problem that they were unable to identify during the manual inspection.

4 - Data Analysis. The previous steps of our procedure produced the following data: (i) a list of OIs identified by our approach; and (ii) a list of problems identified by the system analyst with or without the aid of our approach. Our aim is that our approach detects all and only real problems (i.e., all OIs are real problems and all real problems are identified as OIs). This can be seen as a *classification problem*, and thus the effectiveness of our approach can be measured using the metrics widely used in the context of information retrieval of *precision* and *recall* [13], whose formulas are shown below:

$$Precision = \frac{true\ positives}{true\ positives + false\ positives} \tag{1}$$

$$Recall = \frac{true\ positives}{true\ positives + false\ negatives} \tag{2}$$

where *true positives* are OIs that correspond to real problems; *false positives* are OIs that are not real problems; and *false negatives* are real problems not identified as OIs.

3.2 Target System

The UC descriptions we used in our study are part of the analysis documentation of an industrial software project. This project involves the development of a typical system to manage and sell products, and include functional requirements such as adding new products, changing product information, creating sale orders, and releasing products in stock. Because our study procedure involves the manual analysis of UC descriptions, we selected a subset of all available UCs, choosing those that are not trivial, involving basic and alternative flows. The selected UC descriptions were written in English and described actions performed by actors (e.g. user, system, database, etc.) to achieve a particular state of the system or perform a specific operation. The UC set used has an average of ten sequential steps and five alternative branches.

Based on the Use Case Points Method [4], the UCs used in the study were evaluated between *average* and *complex* because of the number of transactions (between 4 and 7 or more than 7) and the type of actors involved (many complex actors) in their descriptions. Figure 12 shows a fragment of a full textual description of a UC along with two examples of *OIs* found.

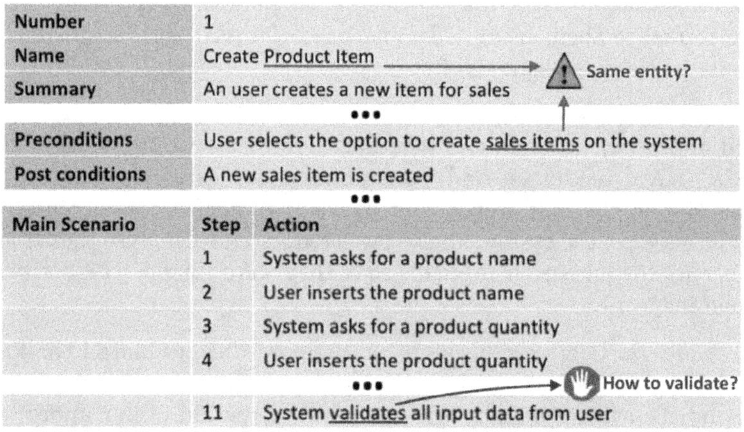
Number
1
Name
Create Product Item
Same entity?
Summary
An user creates a new item for sales
•••
Preconditions
User selects the option to create sales items on the system
Post conditions
A new sales item is created
•••
Main Scenario
Step
Action
1
System asks for a product name
2
User inserts the product name
3
System asks for a product quantity
4
User inserts the product quantity
•••
How to validate?
11
System validates all input data from user

Fig. 12. Example of *Open Issues* found in the Use Case textual description.

In the example, the first OI identified is related to the names of the entities involved in the use case. The first step of the analysis is to identify all existing entities in the description, and in this situation, since there are three different names for an entity which is apparently the same, problems such as uncertainty, inaccuracy or ambiguity may appear in system development. This OI, classified as a warning, has not been confirmed as real problem because *Product Item* and *Sales Item* have different semantics in the system and are indeed treated as different entities. The other OI found is more serious, because there is an effect that is not fully defined in the use case. It is unclear what means the action of *validate* the data. Validation can be related to the existence of the data in the database, or if the data entered are compatible with the expected type (numbers, words, dates, etc.), or check the correctness according to some oracle, etc. The definition of this effect possibly will require a project team's decision, so it is rated at a higher level of severity. In this case, the problem was confirmed, since the analyst was not sure how this validation would be performed. These two situations exemplify the review process applied to the entire set of UCs used in this study. We do not provide any further details about our target system and its UCs due to a confidentiality agreement.

4 Results and Discussion

As result of the execution of the procedure described in the previous section, we collected the data needed to answer our research questions. The system analyst, after revising the original UCs, reported that they had no problems whatsoever. Hence, from the perspective of the analyst who created the UCs, they were correct. However, after applying our approach to these UCs, 32 OIs were identified across the 5 UCs, which gives an average of 6.4 OIs per UC. This is an expressive number, since the system analyst stated that the UCs did not contain any problem. In order to verify whether the identified OIs were false alarms (false positives), the system analyst was asked to check each one of them. Out of the 32 OIs, 24 were pointed out by the analyst as real problems and, consequently, only 8 of the identified OIs were false positives.

Table 6 presents our results in detail. It shows the number of OIs found in each UC (columns #OI) and how many of these OIs were confirmed as real problems (columns #P). The rows show the number of detected OIs with respect to their level of severity (yellow, orange, or red), according to our previously introduced classification. The table also presents the total number of detected OIs of each type and the total number of real problems considering all the 5 UCs.

Table 6. Study results

OI	UC 1		UC 2		UC 3		UC 4		UC 5		Total	
Type	**#OI**	**#P**	**#OI**	**#P**	**#OI**	**#P**	**#OI**	**#P**	**#OI**	**#P**	**#OI**	**#P**
⚠	3	2	4	2	2	1	4	2	1	0	14	7
?	1	1	1	1	1	1	0	0	2	2	5	5
✋	3	3	1	1	3	2	3	3	3	3	13	12
Total	**7**	**6**	**6**	**4**	**6**	**4**	**7**	**5**	**6**	**5**	**32**	**24**

Legend: UC - Use Case; OI - Open Issue; P - Problem.

The results were then analyzed according to the selected metrics. Because the system analyst was unable to identify any problem without support (i.e., to the best of the analyst's knowledge there was no error in the analyzed UCs), the number of problems not identified by our approach was equals to 0, leading to *recall* = 1.0. However, this result is possibly misleading, as there might be problems not identified by both the system analyst and our approach. This is an indication that even for a stakeholder (such as a system analyst, or a user) who knows well the system domain, it is not a trivial task to identify problems. A possible reason for this is the fact that this knowledge of the domain causes the stakeholder to understand different names as synonyms as part of this domain-specific knowledge and overlook omissions in the UCs because they believe some information is obvious. This is an evidence that support (e.g. techniques or tools) for the revision process plays a key role in identifying existing UC problems. As for the Precision, the obtained value was 0.75 (24 true positives and 8 false positives) — that is, 75 % of the OIs identified by our GT-based approach were real problems in the UC descriptions. Most of the identified issues were actual problems whereas most of our false alarms (7 of 8) are in the less critical categories. This means that most of the identified issues not only were real errors not detected by the person who created the UCs, but also revealed problems that required attention from the analyst as they could lead to serious consequences in the actual system.

By analyzing OIs not identified as problems, we observed that 6 of them were not necessarily classified as a false positive by the system analyst. They preferred to leave such issues as they were and postpone changes to future design decisions, considering that they alone could not decide what was the best approach to tackle those issues. The other 2 OIs found, confirmed as false positives by the system analyst, were related to words (names or concepts) used in the specification and have been identified as incompleteness or ambiguities due to lack of knowledge of the modeler about the problem domain and the internal processes of the company. Considering these results, it was concluded that, in most cases, our analysis helped the system analyst, even when an OI did not cause an immediate UC fix, but showed issues that might be considered in future phases of the project. These observations will be used as input for a refinement of the steps proposed to formalize UCs using GTs, in order to create a tool-assisted method to support the UC reviewing process.

Note that OIs were identified without the intervention of any stakeholder. The only provided input was the software documentation in the form of UC descriptions and the output was a checklist with OIs to be revised. For the system analyst, this has great value because the detected problems can be resolved not only at the UC level, but also at the design and implementation levels, as they are performed based on UCs. More importantly, had these problems been detected before the design and implementation, when they should have, development costs could have been potentially reduced.

5 Threats to Validity

When planning and conducting our study, we carefully considered validity concerns. This section discusses the main threats we identified to validate this study and how we mitigated them.

Internal Validity. The main threat to internal validity of this study was the selection of a person responsible for performing the modeling of UCs in the formalism of graphs. Being a formalism mainly used in the sub-area of formal methods, it is difficult to find professionals working on software projects in industry with in-depth knowledge of graph transformations. However, one of our intentions with this work was to show that, correctly following the steps of our strategy, the modeler does not need a deep understanding of the formalism. Moreover, we used the AGG tool to automate the analyses of the generated model and provide a graphical interface for the manipulation of graphs.

Construct Validity. There are different ways of modeling a system through the formalism of graphs that can produce some threats to construct validity. The modeler may not follow correctly the modeling steps, being influenced by their prior knowledge about the formalism. This means that they could change the way of building the model based on their own previous knowledge. Consequently, we cannot guarantee they will obtain similar results to those presented in this work. The same applies when they have an advanced knowledge of the problem domain, because the modeler can insert information in the model that is not documented in the software artifact, hiding a possible omission of information in the UC description. For these reasons, we proposed a roadmap, step by step, on how to model UCs as GTs, for both beginners and experts users.

Conclusion Validity. As the main threat to validity of the conclusion of our study we also highlight potential problems in the generation of the formal model. Besides different forms of modeling and the issue that the modeler may be influenced by their experience or prior knowledge of the problem domain, the modeler may build a model inconsistent with the initial documentation due to errors during the modeling process. Once again, our step-by-step modeling process should be followed to prevent the creation of a model that is not consistent with the UC. Moreover, the tool-supported verifications can also detect some modeling errors, as shown in Tables 2, 3 and 4, thus reducing the risk of this threat.

External Validity. The main threat to the external validity was the selection of artifacts on which we based our study. We did not use any criteria to select either the project or the system analyst who participated of our study. As a consequence of this, the project that was made available for us may not be a representative sample of a large set of software development projects. We were aware of this threat during the study. However, we opted for randomly choosing artifacts to support the applicability of our strategy in different scenarios. This way, we guaranteed that we were not selecting UCs that would be more tailored for our approach. We also believe that obtaining good results even in a situation of a random choice of UCs gives greater confidence on our process.

6 Related Work

Some authors have developed approaches for translating UCs to well-known formalisms, such as LTS [16], Petri Nets [20], and FSM [9,17]. Unlike these formalisms, a GT model is data-driven, hence the focus is on the manipulation of data inside the system. We do not need to explicitly determine the control flow unless it is necessary to guarantee data consistency. Considering other approaches that formalize UCs using a GT model, there are two closest to ours. The approach presented in [21] allows the simulation of the execution of the system but do not report the use of any type of analysis, which, in our opinion, reduces the advantage of having a formal model. The work described in [6] considers analyses such as critical pairs and dependencies involving multiple UCs and provides some ideas on the interpretation of the results. However, we propose a more structured way of providing diagnostic feedback about single UCs, which serves as a guide to point out the possible errors as well as their severity level. As problems in individual UCs can affect the inter-UCs behavior, we chose to initially study how to improve each UC and then move on to the study of how to apply similar ideas for inter-UC formalization and analysis.

Although we could find similar approaches regarding the formalization of UCs as GT models, we could not find any description of an empirical study as the one described in this paper. We believe that this type of study is important not only to provide confidence on the proposed approach, encouraging us to develop it further in terms of its formalization as a validation and verification process, but also to allow us to quantify how good it is. This type of result is also important from the stakeholders' point of view, as they can see in practice and numerically the benefits of applying formal methods to a usually informal process. Moreover, unlike most of the other approaches, our work focuses on helping the developers to construct the formal model by the definition of a systematic translation.

7 Conclusions and Future Work

In this paper, was investigated the suitability of GT as a formal basis for UC description and improvement. Was defined an outline of a translation process from UCs to GTs in a step-by-step manner, which describes how to use an existing tool to analyze the generated model and diagnose real and potential problems. The detection of such problems may cause changes to the UCs and trigger a new round of analyses, incrementally and iteratively improving the initial specification. The process also generates a formal model that can be used for further analyses. The approach was evaluated through an experiment with real software artifacts, where it was possible to detect existing errors, allowing an improvement on the original UCs.

Making a general analysis of the experiment, the results were considered promising, since it was possible to identify a large number of real problems based on a documentation that was produced at an early stage of a software development. Considering the proposed strategy, was observed the need for further automation of the process, which is the most immediate planned future work.

In this paper, was discussed the application of the approach to one UC at a time. Even though this has already shown benefits regarding the software development process, inter-UC analyses are currently being implemented as well as the appropriate diagnostic feedback. Within the same model frame, other types of validation and verification techniques on GT models, such as test case generation, model checking, and theorem proving are also subject of current work. These techniques will be incorporated into a comprehensive methodology for software quality improvement targeting other types of errors. We also plan to study how changes in the UCs could be handled by our approach and how we could reduce the impact and cost of changes by identifying which parts are effectively affected and which analyses are actually required.

Also, a tool was designed, which is already in the early stages of development, in order to automate the first steps of the methodology (between steps 1 and 5) to help the analyst to build a formal model through the graph formalism. This tool aims to help produce the model data in a format acceptable by the AGG tool, responsible for the computational analysis of the graphs.

Finally, note that, although was not presented any new formal method or verification technique here, a considerable amount of expertise in formal methods was required to define the OIs: they are meant to bridge the gap between the informal and formal worlds. We believe that this type of work is crucial towards the industrial adoption of formal methods.

References

1. Alagar, V.S., Periyasamy, K.: Specification of Software Systems. Texts in Computer Science, 2nd edn. Springer, London (2011)
2. Basili, V., Caldiera, C., Rombach, H.: Goal question metric paradigm. In: Marciniak, J.J. (ed.) Encyclopedia of Software Engineering, vol. 1. Wiley, New York (1994)
3. Cockburn, A.: Writing Effective Use Cases, 1st edn. Addison-Wesley Longman Publishing Co., Inc., Boston (2000)
4. Diev, S.: Use cases modeling and software estimation: applying use case points. SIGSOFT Softw. Eng. Notes **31**(6), 1–4 (2006)
5. Ehrig, H., Engels, G., Kreowski, H.J., Rozenberg, G. (eds.): Applications, languages, and tools, vol. 2. World Scientific, River Edge (1999)
6. Hausmann, J.H., Heckel, R., Taentzer, G.: Detection of conflicting functional requirements in a use case-driven approach: a static analysis technique based on graph transformation. In: Proceedings of the 24th ICSE, pp. 105–115 (2002)
7. Hurlbut, R.R.: A survey of approaches for describing and formalizing use cases. Technical report XPT-TR-97-03, Expertech, Ltd. (1997)
8. Jin, N., Yang, J.: An approach of inconsistency verification of use case in XML and the model of verification tool. In: Proceedings of MINES 2010, pp. 757–761 (2010)
9. Klimek, R., Szwed, P.: Formal analysis of use case diagrams. Comp. Sci. **11**, 115–131 (2010)
10. Köters, G., werner Six, H., Winter, M.: Validation and verification of use cases and class models. In: Proceedings of the 6th REFSQ (2001)
11. Myers, G., Sandler, C., Badgett, T.: The Art of Software Testing. ITPro Collection. Wiley, New York (2011)
12. Patton, R.: Software Testing, vol. 408, 2nd edn. Sams Publishing, Indianapolis (2005)

13. Powers, D.M.: Evaluation: from precision, recall and F-factor to ROC, informedness, markedness & correlation. Technical Report SIE-07-001, FUSA (2007)
14. Rozenberg, G. (ed.): Foundations, vol. 1. World Scientific, River Edge (1997)
15. Shen, W., Liu, S.: Formalization, testing and execution of a use case diagram. In: Dong, J.S., Woodcock, J. (eds.) ICFEM 2003. LNCS, vol. 2885, pp. 68–85. Springer, Heidelberg (2003)
16. Sinnig, D., Chalin, P., Khendek, F.: LTS semantics for use case models. In: Proceedings of the ACM SAC, pp. 365–370. ACM (2009)
17. Sinnig, D., Chalin, P., Khendek, F.: Use case and task models: an integrated development methodology and its formal foundation. ACM ToSEM **22** (2013)
18. Taentzer, G.: AGG: a tool environment for algebraic graph transformation. In: Münch, M., Nagl, M. (eds.) AGTIVE 1999. LNCS, vol. 1779, pp. 481–488. Springer, Heidelberg (2000)
19. Wohlin, C., Runeson, P., Höst, M., Ohlsson, M.C., Regnell, B., Wesslén, A.: Experimentation in Software Engineering. Springer, Berlin (2012)
20. Zhao, J., Duan, Z.: Verification of use case with petri nets in requirement analysis. In: Gervasi, O., Taniar, D., Murgante, B., Laganà, A., Mun, Y., Gavrilova, M.L. (eds.) ICCSA 2009, Part II. LNCS, vol. 5593, pp. 29–42. Springer, Heidelberg (2009)
21. Ziemann, P., Hǎv̌lscher, K., Gogolla, M.: From UML models to graph transformation systems. ENTCS **127**(4), 17–33 (2005)

A Full Operational Semantics for Asynchronous Relational Networks

Ignacio Vissani[1(✉)], Carlos Gustavo Lopez Pombo[1,2], Ionuţ Ţuţu[3,4], and José Luiz Fiadeiro[3]

[1] Departamento de Computación, Universidad de Buenos Aires, Buenos Aires, Argentina
{ivissani,clpombo}@dc.uba.ar
[2] Consejo Nacional de Investigaciones Científicas y Tecnológicas, Buenos Aires, Argentina
[3] Department of Computer Science, Royal Holloway University of London, Surrey, UK
ittutu@gmail.com, Jose.Fiadeiro@rhul.ac.uk
[4] Institute of Mathematics of the Romanian Academy, Research Group of the Project ID-3-0439, Bucharest, Romania

Abstract. Service-oriented computing is a new paradigm where applications run over global computational networks and are formed by services discovered and bound at run-time through the intervention of a middleware. *Asynchronous Relational Nets* (ARNs) were presented by Fiadeiro and Lopes with the aim of formalising the elements of an interface theory for service-oriented software designs. The semantics of ARNs was originally given in terms of sequences of sets of actions corresponding to the behaviour of the service. Later, they were given an institution-based semantics where signatures are ARNs and models are morphisms into ground networks, that have no dependencies on external services.

In this work, we propose a full operational semantics capable of reflecting the dynamic nature of service execution by making explicit the reconfigurations that take place at run-time as the result of the discovery and binding of required services. This provides us a refined view of the execution of ARNs based upon which a specialized variant of linear temporal logic can be used to express, and even to verify through standard model-checking techniques, properties concerning the behaviour of ARNs that are more complex than those considered before.

1 Introduction and Motivation

In the context of global ubiquitous computing, the structure of software systems is becoming more and more dynamic as applications need to be able to respond and adapt to changes in the environment in which they operate. For instance, the new paradigm of *Service-Oriented Computing* (SOC) supports a new generation of software applications that run over globally available computational

This work has been supported by the European Union Seventh Framework Programme under grant agreement no. 295261 (MEALS).

M. Codescu et al. (Eds.): WADT 2014, LNCS 9463, pp. 131–150, 2015.
DOI: 10.1007/978-3-319-28114-8_8

and network infrastructures where they can procure services on the fly (subject to a negotiation of *Service Level Agreements*, or SLAs for short) and bind to them so that, collectively, they can fulfil given business goals [12]. There is no control as to the nature of the components that an application can bind to. In particular, development no longer takes place in a top-down process in which subsystems are developed and integrated by skilled engineers: in SOC, discovery and binding are performed by middleware.

Asynchronous Relational Networks (ARNs) were presented by Fiadeiro and Lopes in [13] with the aim of formalising the elements of an interface theory for service-oriented software designs. ARNs are a formal orchestration model based on hypergraphs whose hyperedges are interpreted either as processes or as communication channels. The nodes (or points) that are only adjacent to process hyperedges are called *provides-points*, while those adjacent only to communication hyperedges are called *requires-points*. The rationale behind this separation is that a provides-point is the interface through which a service exports its functionality, while a requires-point is the interface through which an activity expects certain service to provide a functionality. The composition of ARNs (i.e., the binding mechanism of services) is obtained by "fusing" provides-points and requires-points, subject to a certain compliance check between the contract associated to them. For example, in [22] the binding is subject to a (semantic) entailment relation between theories over *linear temporal logic* [19], which are attached to the provides- and the requires-points of the considered networks.

Providing semantics to ARNs requires to carefully combine different elements intervening in the rationale behind the formalism and its intended behaviour. In their first definition, ARNs were given semantics in terms of infinite sequences of sets of actions, which capture the behaviour of the service. In this presentation, the behavioural description was given in terms of linear temporal logic theory presentations [13]. A more modern (and also more operational) presentation of the semantics of ARNs, the one on which we rely in this article, resorts to automata on infinite objects whose inputs consist of sequences of sets of actions (see [22]), as defined in the original semantics of ARNs. Under this formalism, both types of hyperedges are labelled with Muller automata; in the case of process hyperedges, the automata formalise the computation carried out by that particular service, while in the case of communication hyperedges, the automata represent the orchestrator that syncs the behaviour of the participants in the communication process. The behaviour of the system is then obtained as the composition of the Muller automata associated to both computation and communication hyperedges. Finally, the reconfiguration of networks (realized through the discovery and binding of services) is defined by considering an institutional framework in which signatures are ARNs and models are morphisms into ground ARNs, which have no dependencies on external services (see, e.g., [22] for a more in depth presentation of this semantics).

Under the above-mentioned consideration, the operational semantics of ARNs (as a set of execution traces) is based on the fact that a network can be reconfigured until all its external dependencies (captured by requires-points) are fulfilled, i.e., the original network admits a morphism to a ground ARN. In our work, we consider that semantics is assigned modulo a given repository of services, forcing us to drop the assumption that given an ARN it is possible to find a ground network to which the former has a morphism. Regarding previous works, we believe that this approach results in a more realistic executing environment where the potential satisfaction of requirements is limited by the services registered in a repository, and not by the entire universe of possible services.

The aim of this work is to provide a trace-based operational semantics for service-oriented software designs reflecting the true dynamic nature of run-time discovery and binding of services. This is done by making the reconfiguration of an activity an observable event of its behaviour. In SOC, the reconfiguration events are triggered by particular actions associated with a requires-point; at that moment, the middleware has to procure a service that meets the requirements of the activity from an already known repository of services. From this perspective our proposal is to define execution traces where actions can be either

- internal actions of the activity: actions that are not associated with requires-points, thus executable without the need for reconfiguring the activity, or
- reconfiguration actions: actions that are associated with a requires-point, thus triggering the reconfiguration of the system by means of the discovery and binding of a service providing that computation.

Summarising, the main contributions of this paper are: (1) we provide a trace-based operational semantics for ARNs reflecting both internal transitions taking place in any of the services already intervening in the computation and dynamic reconfiguration actions resulting from the process by binding the provides-point of ARNs taken from the repository to its require-points, and (2) we provide support for defining a model-checking technique that can enable the automatic analysis of linear temporal logic properties of activities.

In this way, our work departs from previous approaches to dynamic reconfiguration in the context of service-oriented computing, such as [20], which reasons about functional behaviour and control concerns in a framework based on first-order logic, [6], which relies on typed graph-transformation techniques implemented in Alloy [15] and Maude [7,10], which makes use of graph grammars as a formal framework for dealing with dynamicity, and [8,14], which proposes architectural design rewriting as a term-rewriting-based approach to the development and reconfiguration of software architectures. A survey of these general logic-, graph-, or rewriting-based formalisms can be found in [4].

The article is organised as follows. In Sect. 2 we recall the preliminary notions needed for our work. In Sect. 3 we give appropriate definitions for providing operational semantics for ARNs based on a (quasi) automaton *generated* by a repository and on the traces accepted by it. We also provide a variant of Linear Temporal Logic (in Sect. 4) that is suitable for defining and checking properties related to the execution of activities. As a running example, we gradually introduce the details of travel-agent scenario, which we use to illustrate the concepts presented in the Sects. 3 and 4. Finally in Sect. 5 we draw some conclusions and discuss further lines of research.

2 Preliminary Definitions

In this section we present the preliminary definitions we use throughout this work. We assume the reader has a nodding acquaintance of the basic definitions of category theory. Most of the definitions needed throughout the forthcoming sections can be found in [2,11,17]. For hypergraph terminology, notation and definitions, the reader is pointed to [1,9], while for automata on infinite sequences we suggest [18,21].

Definition 1 (Muller automaton). *The category* $\mathbb{MA}$ *of (action-based) Muller automata (see, e.g. [22]) is defined as follows:*
The objects *of* $\mathbb{MA}$ *are pairs* $\langle A, \Lambda \rangle$ *consisting of a set* A *of* actions *and a* Muller automaton $\Lambda = \langle Q, 2^A, \Delta, I, \mathcal{F} \rangle$ *over the alphabet* 2^A*, where*

- Q *is the set of* states *of* Λ*,*
- $\Delta \subseteq Q \times 2^A \times Q$ *is the* transition relation *of* Λ*, with transitions* $(p, \iota, q) \in \Delta$ *usually denoted by* $p \xrightarrow{\iota} q$*,*
- $I \subseteq Q$ *is the set of* initial states *of* Λ*, and*
- $\mathcal{F} \subseteq 2^Q$ *is the set of* final-state sets *of* Λ*.*

For every pair of Muller automata $\langle A, \Lambda \rangle$ *and* $\langle A', \Lambda' \rangle$*, with* $\Lambda = \langle Q, 2^A, \Delta, I, \mathcal{F} \rangle$ *and* $\Lambda' = \langle Q', 2^{A'}, \Delta', I', \mathcal{F}' \rangle$*, an* $\mathbb{MA}$-morphism $\langle \sigma, h \rangle \colon \langle A, \Lambda \rangle \to \langle A', \Lambda' \rangle$ *consists of functions* $\sigma \colon A \to A'$ *and* $h \colon Q' \to Q$ *such that* $(h(p'), \sigma^{-1}(\iota'), h(q')) \in \Delta$ *whenever* $(p', \iota', q') \in \Delta'$*,* $h(I') \subseteq I$*, and* $h(\mathcal{F}') \subseteq \mathcal{F}$*.*
The composition *of* $\mathbb{MA}$*-morphisms is defined componentwise.*

As we mentioned before, in this work we focus on providing semantics to service-oriented software artefacts. To do that, we resort to the formal language of *asynchronous relational nets* (see, e.g., [13]). The intuition behind the definition is that ARNs are hypergraphs where the hyperedges are divided in two classes: computation hyperedges and communication hyperedges. Computation hyperedges represent processes, while communication hyperedges represent communication channels. Hypergraph nodes (also called points) are labelled with *ports*, i.e., with structured sets $M = M^- \cup M^+$ of *publication* (M^-) and *delivery messages* (M^+),[1] along the lines of [3,5]. At the same time, hyperedges are labelled with Muller automata; thus, both processes and communication channels have an associated behaviour given by their corresponding automata, which interact through (messages defined by) the ports labelling their connected points.

The following definitions formalise the manner in which the computation and communication units are structured to interact with each other.

Definition 2 (Process). *A* process $\langle \gamma, \Lambda \rangle$ *consists of a set* γ *of pairwise disjoint ports and a Muller automaton* Λ *over the set of actions* $A_\gamma = \bigcup_{M \in \gamma} A_M$*, where* $A_M = \{m! \mid m \in M^-\} \cup \{m¡ \mid m \in M^+\}$*.*

[1] Formally, we can define ports as sets M of messages together with a function $M \to \{-, +\}$ that assigns a polarity to every message.

As an example, Fig. 1 (a) depicts a process $\langle\gamma_{\mathsf{TA}}, \Lambda_{\mathsf{TA}}\rangle$ where $\gamma_{\mathsf{TA}} = \{\mathsf{TA}_0, \mathsf{TA}_1, \mathsf{TA}_2\}$ and Λ_{TA} is the automaton presented in Fig. 1 (b). The travel agent is meant to provide hotel and/or flight bookings in the local currency of the customers. Accomplishing this task requires two different interactions to take place: on one hand, the communication with hotel-accommodation providers and with flight-tickets providers, and on the other hand, the communication with a currency-converter provider. In order for the composition of the automata developed along our example to behave well, we need that every automaton is able to stay in any state indefinitely. This behaviour is achieved by forcing every state to have a self-loop labelled with the emptyset. With the purpose of easing the figures we avoid drawing these self-loops. The reader should still understand the descriptions of the automata as if there was a self-loop transition, labelled with the empty set, for every state.

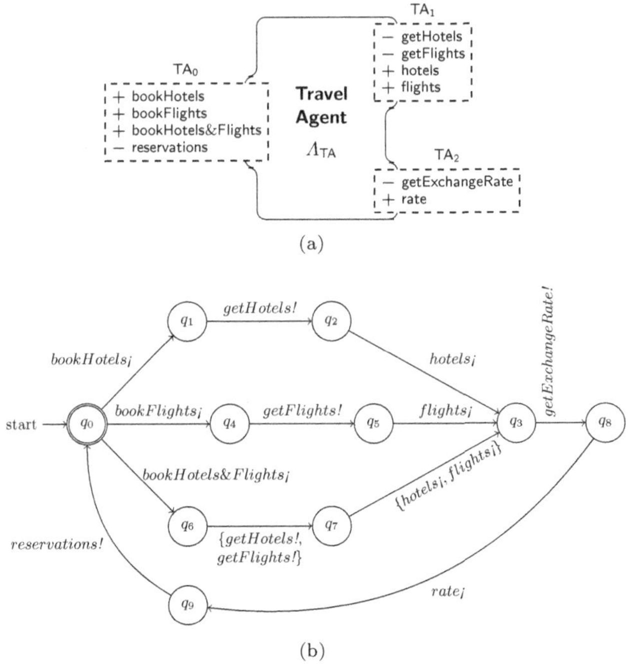

Fig. 1. The TravelAgent process together with its automaton Λ_{TA}

Definition 3 (Connection). *Let γ be a set of pairwise disjoint ports. A connection $\langle M, \mu, \Lambda\rangle$ between the ports of γ consists of a set M of messages, a partial attachment injection $\mu_i \colon M \rightharpoonup M_i$ for each port $M_i \in \gamma$, and a Muller automaton Λ over $A_M = \{m! \mid m \in M\} \cup \{m¡ \mid m \in M\}$ such that*

$$(a) \quad M = \bigcup_{M_i \in \gamma} \mathrm{dom}(\mu_i) \quad \textit{and} \quad (b) \quad \mu_i^{-1}(M_i^{\mp}) \subseteq \bigcup_{M_j \in \gamma \setminus \{M_i\}} \mu_j^{-1}(M_j^{\pm}).$$

In Fig. 2 (a) a connection C_0 is shown whose set of messages is the union of the messages of the ports $\mathsf{TA}_1, \mathsf{H}_0, \mathsf{F}_0$ and the family of mappings μ is formed by the trivial identity mapping. In Fig. 2 (b) the automaton Λ_{C_0} that describes the behaviour of the communication channel is shown. This connection just delivers every published message. Nevertheless it imposes some restrictions to the sequences of messages that can be delivered. For example notice that, after the message getHotels of TA_1 is received (and delivered), only the message hotels of H_0 is accepted for delivery.

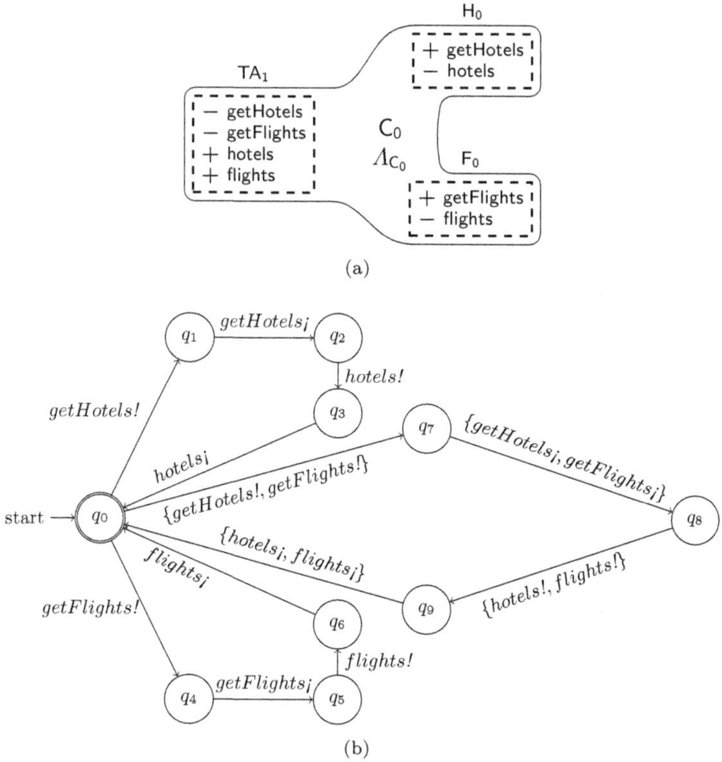

Fig. 2. The C_0 connection

With these elements we can now define asynchronous relational networks.

Definition 4 (Asynchronous Relational Net [22]). *An* asynchronous relational net $\alpha = \langle X, P, C, \gamma, M, \mu, \Lambda \rangle$ *consists of*

- *a hypergraph* $\langle X, E \rangle$*, where* X *is a (finite) set of* points *and* $E = P \cup C$ *is a set of* hyperedges *(non-empty subsets of* X*) partitioned into* computation hyperedges $p \in P$ *and* communication hyperedges $c \in C$ *such that no adjacent hyperedges belong to the same partition, and*

- *three labelling functions that assign (a) a port M_x to each point $x \in X$, (b) a process $\langle \gamma_p, \Lambda_p \rangle$ to each hyperedge $p \in P$, and (c) a connection $\langle M_c, \mu_c, \Lambda_c \rangle$ to each hyperedge $c \in C$.*

Definition 5 (Morphism of ARNs). *A* morphism $\delta \colon \alpha \to \alpha'$ *between two ARNs* $\alpha = \langle X, P, C, \gamma, M, \mu, \Lambda \rangle$ *and* $\alpha' = \langle X', P', C', \gamma', M', \mu', \Lambda' \rangle$ *consists of*

- *an injective map* $\delta \colon X \to X'$ *such that* $\delta(P) \subseteq P'$ *and* $\delta(C) \subseteq C'$*, that is an injective homomorphism between the underlying hypergraphs of* α *and* α' *that preserves the computation and communication hyperedges, and*
- *a family of polarity-preserving injections* $\delta^{pt}_x \colon M_x \to M'_{\delta(x)}$*, for* $x \in X$*,*

such that

- *for every point* $x \in \bigcup P$, $\delta^{pt}_x = 1_{M_x}$*,*
- *for every computation hyperedge* $p \in P$, $\Lambda_p = \Lambda'_{\delta(p)}$*, and*
- *for every communication hyperedge* $c \in C$, $M_c = M'_{\delta(c)}$, $\Lambda_c = \Lambda'_{\delta(c)}$ *and, for every point* $x \in \gamma_c$, $\mu_{c,x} ; \delta^{pt}_x = \mu'_{\delta(c),\delta(x)}$.

ARNs together with morphisms of ARNs form a category, denoted $\mathbb{ARN}$, in which the composition is defined component-wise, and left and right identities are given by morphisms whose components are identity functions.

Intuitively, an ARN is a hypergraph for which some of the hyperedges (process hyperedges) formalise computations as Muller automata communicating through ports (identified with nodes of the hypergraph) over a fixed language of actions. Note that the communication between computational units is not established directly but mediated by a communication hyperedge; the other kind of hyperedge which use Muller automata to formalise communication channels.

In order to define service modules, repositories, and activities, we need to distinguish between two types of interaction-points, i.e. of points that are not incident with both computation and communication hyperedges.

Definition 6 (Requires- and provides-point). *A* requires-point *of an ARN is a point that is incident only with a communication hyperedge. Similarly, a* provides-point *is a point incident only with a computation hyperedge.*

Definition 7 (Service repository). *A* service repository *is just a set* $\mathcal{R}$ *of* service modules*, that is of triples* $\langle P, \alpha, R \rangle$*, also written* $P \xleftarrow[\alpha]{} R$*, where* α *is an ARN,* P *is a provides-point of* α*, and* R *is a finite set of requires-points of* α*.*

Definition 8 (Activity). *An* activity *is a pair* $\langle \alpha, R \rangle$*, also denoted* $\vdash_{\alpha} R$*, such that* α *is an ARN and* R *is a finite set of requires-points of* α*.*

The previous definitions formalise the idea of a service-oriented software artefact as an activity whose computational requirements are modelled by "dangling" connections, and that do not pursue the provision of any service to other computational unit, modelled as the absence of provides points. Figure 3 depicts a TravelClient activity with a single requires-point through which this activity

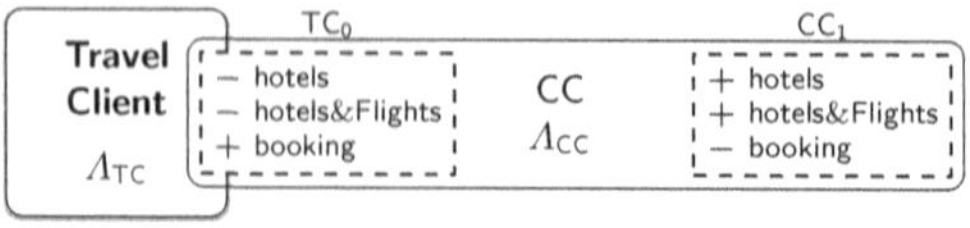

(a) The TravelClient process.

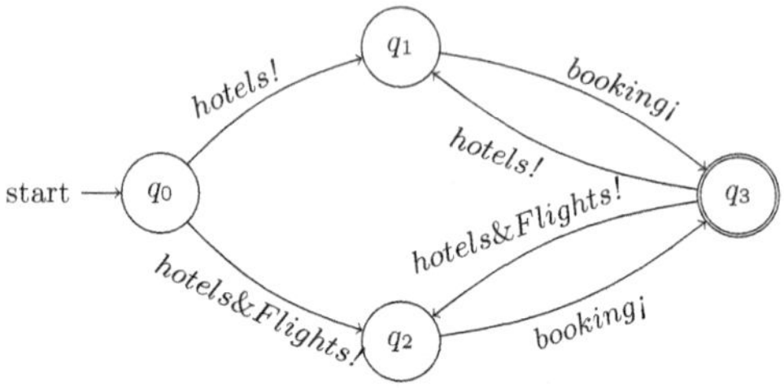

(b) The automaton Λ_{TC}

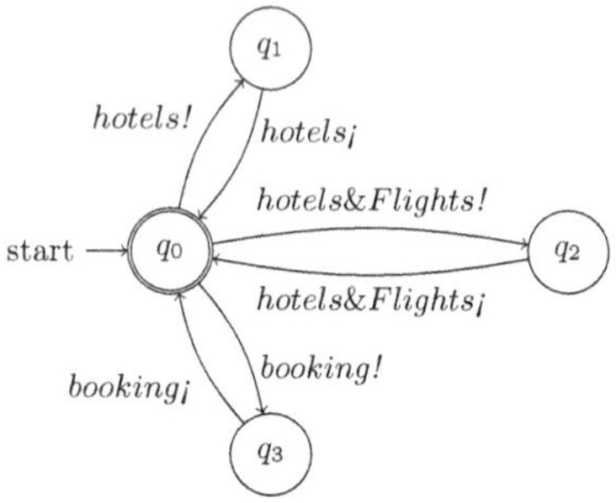

(c) The automaton Λ_{CC}

Fig. 3. The TravelClient activity.

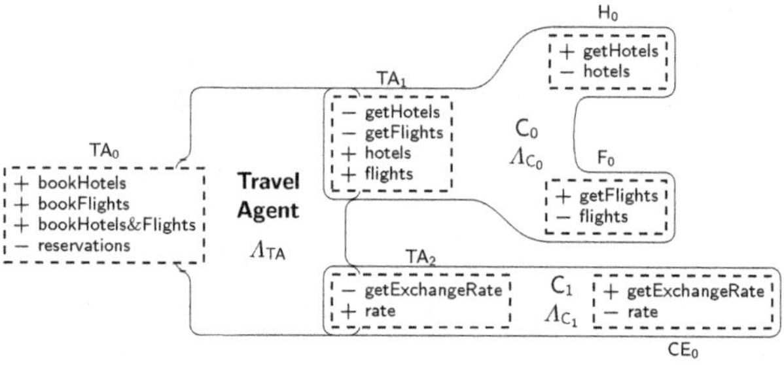

(a) The TravelAgent service.

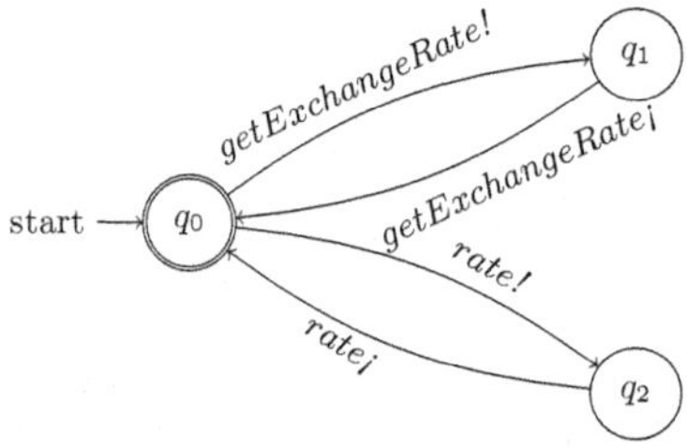

(b) The automaton Λ_{C_1}.

Fig. 4. The TravelAgent service module.

can ask either for hotels or hotels and flights reservations. As we will show in the forthcoming sections, requires-points act as the ports to which the provides-points of services are bound in order to fulfil these requirements.

Turning a process into a service available for discovery and binding requires, as we mentioned in the previous definitions, the declaration of the communication channels that will be used to connect to other services. In the case of TravelAgent, three services are required to execute, communicating over two different communication channels. On one of them the process interacts with accommodation providers and flight tickets providers, while through the other the process will obtain exchange rates to be able to show the options to the customer in its local currency. In some sense, TravelAgent provides the ability to coherently combine these three services in order to offer a richer experience. It should then be clear that whenever the TravelAgent is asked for hotels and flights reservations, it will require both services in order to fulfil its task, plus the service for currency exchange conversion. Figure 4 (a) shows the TravelAgent service obtained by attaching the communication channels to two of the ports defined by the TravelAgent process, resulting in a network with three requires-points. The Fig. 4 (b) shows the automaton for the connection C_1.

3 Operational Semantics for ARNs

In this section we present the main contribution of the paper, being a full operational semantics for activities executing with respect to a given repository. To do this, we introduce two different kinds of transitions for activities: (1) item internal transitions, those resulting from the execution of a certain set of actions by the automata that synchronise over them, and (2) reconfiguration transitions, the ones resulting from the need of executing a set of actions on a port of a communication hyperedge. Then, runs (on given traces) are legal infinite sequences of states related by appropriate transitions.

Definition 9 (Alphabet of an ARN). *The* alphabet associated with an ARN *α is the vertex A_α of the colimit $\xi\colon D_\alpha \Rightarrow A_\alpha$ of the functor $D_\alpha\colon \mathbb{J}_\alpha \to$* Set, *where*

- *$\mathbb{J}_\alpha$ is the preordered category whose objects are points $x \in X$, hyperedges $e \in P \cup C$, or "attachments" $\langle c, x\rangle$ of α, with $c \in C$ and $x \in \gamma_c$, and whose arrows are given by $\{x \to p \mid p \in P, x \in \gamma_p\}$, for computation hyperedges, and $\{c \leftarrow \langle c, x\rangle \to x \mid c \in C, x \in \gamma_c\}$, for communication hyperedges;*
- *D_α defines the sets of actions associated with the ports, processes and channels, together with the appropriate mappings between them.*

Definition 10 (Automaton of an ARN). *Let $\alpha = \langle X, P, C, \gamma, M, \mu, \Lambda\rangle$ be an ARN and $\langle Q_e, 2^{A_{M_e}}, \Delta_e, I_e, \mathcal{F}_e\rangle$ be the components of Λ_e, for each $e \in P \cup C$. The automaton $\Lambda_\alpha = \langle Q_\alpha, 2^{A_\alpha}, \Delta_\alpha, I_\alpha, \mathcal{F}_\alpha\rangle$* associated with *$\alpha$ is defined as follows:*

$$
\begin{aligned}
&Q_\alpha = \textstyle\prod_{e \in P \cup C} Q_e,\\
&\Delta_\alpha = \{(p, \iota, q) \mid (\pi_e(p), \xi_e^{-1}(\iota), \pi_e(q)) \in \Delta_e \text{ for each } e \in P \cup C\},\\
&I_\alpha = \textstyle\prod_{e \in P \cup C} I_e, \text{ and}\\
&\mathcal{F}_\alpha = \{F \subseteq Q_\alpha \mid \pi_e(F) \in \mathcal{F}_e \text{ for all } e \in P \cup C\},
\end{aligned}
$$

where $\pi_e\colon Q_\alpha \to Q_e$ are the corresponding projections of the product $\prod_{e \in P \cup C} Q_e$.

Fact 1. Under the notations of Definition 10, for every hyperedge e of α, the maps ξ_e and π_e define an $\mathbb{MA}$-morphism $\langle \xi_e, \pi_e\rangle\colon \langle A_{M_e}, \Lambda_e\rangle \to \langle A_\alpha, \Lambda_\alpha\rangle$.

Intuitively, the automaton of an ARN is the automaton resulting from taking the product of the automata of the several components of the ARN. This product is synchronized over the shared alphabet of the components. Notice that the notion of *shared alphabet* is given by the mappings defined in the connections.

Proposition 1. *For every ARN* $\alpha = \langle X, P, C, \gamma, M, \mu, \Lambda \rangle$*, the* $\mathbb{MA}$*-morphisms* $\langle \xi_e, \pi_e \rangle : \langle A_{M_e}, \Lambda_e \rangle \to \langle A_\alpha, \Lambda_\alpha \rangle$ *associated with hyperedges* $e \in P \cup C$ *form colimit injections for the functor* $G_\alpha : \mathbb{J}_\alpha \to \mathbb{MA}$ *that maps*

- *every computation or communication hyperedge* $e \in P \cup C$ *to* $\langle A_{M_e}, \Lambda_e \rangle$ *and*
- *every point* $x \in X$ *(or attachment* $\langle c, x \rangle$*) to* $\langle A_{M_x}, \Lambda_x \rangle$*, where* Λ_x *is the Muller automaton* $\langle \{q\}, 2^{A_{M_x}}, \{(q, \iota, q) \mid \iota \subseteq A_{M_x}\}, \{q\}, \{\{q\}\} \rangle$ *with only one state, which is both initial and final, and with all possible transitions.*

Therefore, both the alphabet and the automaton of an ARN α *are given by the vertex* $\langle A_\alpha, \Lambda_\alpha \rangle$ *of a colimiting cocone of the functor* $G_\alpha : \mathbb{J}_\alpha \to \mathbb{MA}$*.*

The universality property discussed above of the alphabet and of the automaton of an ARN allows us to extend Definition 10 to morphisms of networks.

Corollary 1. *For every morphism of ARNs* $\delta : \alpha \to \alpha'$ *there exists a unique* $\mathbb{MA}$*-morphism* $\langle A_\delta, \_\!\upharpoonright_\delta \rangle : \langle A_\alpha, \Lambda_\alpha \rangle \to \langle A_{\alpha'}, \Lambda_{\alpha'} \rangle$ *such that*

$$(a)\ \ \xi_x ; A_\delta = \xi'_{\delta(x)} \quad \text{and} \quad (b)\ \ (\_\!\upharpoonright_\delta) ; \pi_x = \pi'_{\delta(x)}$$

for every point or hyperedge x *of* α*, where* $\langle \xi_x, \pi_x \rangle$ *and* $\langle \xi'_{x'}, \pi'_{x'} \rangle$ *are components of the colimiting cocones of the functors* $G_\alpha : \mathbb{J}_\alpha \to \mathbb{MA}$ *and* $G_{\alpha'} : \mathbb{J}_{\alpha'} \to \mathbb{MA}$*.*[2]

Operational semantics of ARNs. From a categorical perspective, the uniqueness aspect of Corollary 1 is particularly important in capturing the operational semantics of ARNs in a fully abstract manner: it enables us to describe both automata and morphisms of automata associated with ARNs and morphisms of ARNs through a functor $\mathcal{A} : \mathbb{ARN} \to \mathbb{MA}$ that maps every ARN α to $\langle A_\alpha, \Lambda_\alpha \rangle$ and every morphisms of ARNs $\delta : \alpha \to \alpha'$ to $\langle A_\delta, \_\!\upharpoonright_\delta \rangle$.

3.1 Open Executions of ARNs

In order to formalise open executions of ARNs, i.e. of executions in which not only the states of the underlying automata of ARNs can change as a result of the publication or the delivery of various messages, but also the ARNs themselves through discovery and binding to other networks, we rely on the usual automata-theoretic notions of execution, trace, and run, which we consider over a particular (super-)automaton of ARNs and local states of their underlying automata.

Definition 11. *The "flattened" automaton* $\mathcal{A}^\sharp = \langle Q^\sharp, A^\sharp, \Delta^\sharp, I^\sharp, \mathcal{F}^\sharp \rangle$ *induced by the functor* $\mathcal{A} : \mathbb{ARN} \to \mathbb{MA}$[3] *is defined as follows:*

[2] The definitions of G_α and $G_{\alpha'}$ follow the presentation given in Proposition 1.
[3] Note that $\Lambda^\sharp$ is in fact a quasi-automaton, because its components are proper classes.

$$\begin{aligned}
Q^\sharp &= \{\langle \alpha, q\rangle \mid \alpha \in |\mathbb{ARN}| \textit{ and } q \in Q_\alpha\},\\
A^\sharp &= \{\langle \delta, \iota\rangle \mid \delta\colon \alpha \to \alpha' \textit{ and } \iota \subseteq A_\alpha\},\\
\Delta^\sharp &= \{(\langle \alpha, q\rangle, \langle \delta, \iota\rangle, \langle \alpha', q'\rangle) \mid \delta\colon \alpha \to \alpha' \textit{ and } (q, \iota, q'\restriction_\delta) \in \Delta_\alpha\},\\
I^\sharp &= \{\langle \alpha, q\rangle \mid \alpha \in |\mathbb{ARN}| \textit{ and } q \in I_\alpha\}, \textit{ and}\\
\mathcal{F}^\sharp &= \{\{\langle \alpha, q\rangle \mid q \in F\} \mid \alpha \in |\mathbb{ARN}| \textit{ and } F \in \mathcal{F}_\alpha\}.
\end{aligned}$$

This "flattened" automaton amalgamates in a single structure both the configuration and the state of the system. These two elements are viewed as a pair ⟨ARN, state⟩. Now the transitions in this automaton can represent state changes and structural changes together. In this sense, the "flattened" automaton achieves the goal of giving us a unified view of both aspects of a service oriented system. The construction of this automaton can be seen, from a categorical point of view, as the flattening of the indexed category induced by $\mathcal{A}\colon \mathbb{ARN} \to \mathbb{MA}$.

We recall that a *trace* over a set A of actions is an infinite sequence $\lambda \in (2^A)^\omega$, and that a *run* of a Muller automaton $\Lambda = \langle Q, 2^A, \Delta, I, \mathcal{F}\rangle$ on a trace λ is a sequence of states $\varrho \in Q^\omega$ such that $\varrho(0) \in I$ and $(\varrho(i), \lambda(i), \varrho(i+1)) \in \Delta$ for every $i \in \omega$; together, λ and ϱ form an *execution* of the automaton Λ. An execution $\langle \lambda, \rho\rangle$, or simply the run ϱ, is *successful* if the set of states that occur infinitely often in ϱ, denoted $\mathrm{Inf}(\varrho)$, is a member of $\mathcal{F}$. Furthermore, a trace λ is *accepted* by Λ if and only if there exists a successful run of Λ on λ.

Definition 12 (Open execution of an ARN). *An* open execution *of an ARN α is an execution of $\mathcal{A}^\sharp$ that starts from an initial state of Λ_α, i.e. a sequence*

$$\langle \alpha_0, q_0\rangle \xrightarrow{\delta_0, \iota_0} \langle \alpha_1, q_1\rangle \xrightarrow{\delta_1, \iota_1} \langle \alpha_2, q_2\rangle \xrightarrow{\delta_2, \iota_2} \cdots$$

such that $\alpha_0 = \alpha$, $q_0 \in I_\alpha$ and, for every $i \in \omega$, $\langle \alpha_i, q_i\rangle \xrightarrow{\langle \delta_i, \iota_i\rangle} \langle \alpha_{i+1}, q_{i+1}\rangle$ is a transition in $\Delta^\sharp$. An open execution as above is successful *if it is successful with respect to the automaton $\mathcal{A}^\sharp$, i.e. if there exists $i \in \omega$ such that (a) for all $j \geq i$, $\alpha_j = \alpha_i$, $\delta_j = 1_{\alpha_i}$, and (b) $\{q_j \mid j \geq i\} \in \mathcal{F}_{\alpha_i}$.*

Based on the definition of the transitions of $\mathcal{A}^\sharp$ and on the functoriality of $\mathcal{A}\colon \mathbb{ARN} \to \mathbb{MA}$, it is easy to see that, for every ARN α, every successful open execution of α gives a successful execution of its underlying automaton Λ_α.

Proposition 2. *For every (successful) open execution*

$$\langle \alpha_0, q_0\rangle \xrightarrow{\delta_0, \iota_0} \langle \alpha_1, q_1\rangle \xrightarrow{\delta_1, \iota_1} \langle \alpha_2, q_2\rangle \xrightarrow{\delta_2, \iota_2} \cdots$$

of the quasi-automaton $\mathcal{A}^\sharp$, the infinite sequence

$$q_0 \xrightarrow{\iota_0} q_1\restriction_{\delta_0} \xrightarrow{A_{\delta_0}^{-1}(\iota_1)} q_2\restriction_{\delta_0;\delta_1} \xrightarrow{A_{\delta_0;\delta_1}^{-1}(\iota_2)} \cdots$$

corresponds to a (successful) execution of the automaton Λ_{α_0}.

Note that, since the restrictions imposed to the transitions of $\mathcal{A}^\sharp$ are very weak – more precisely, because there are no constraints on the morphisms of ARN $\delta\colon \alpha \to \alpha'$ that underlie open-transitions $\langle\alpha, q\rangle \xrightarrow{\delta,\iota} \langle\alpha', q'\rangle$ – Proposition 2 cannot be generalised to executions of the automata Λ_{α_i}, for $i > 0$. To address this aspect, we need to take into consideration the fact that, in practice, the reconfigurations of ARNs are actually triggered by certain actions of their alphabet, and that they comply with the general rules of the process of service discovery and binding. Therefore, we need to consider open executions of activities with respect to given service repositories.

3.2 Open Executions of Activities

For the rest of this section we assume that $\mathcal{R}$ is an arbitrary but fixed repository of service modules.

Definition 13. *The* activity (quasi-)automaton $\mathcal{R}^\sharp = \langle Q^{\mathcal{R}}, A^{\mathcal{R}}, \Delta^{\mathcal{R}}, I^{\mathcal{R}}, \mathcal{F}^{\mathcal{R}}\rangle$ *generated by the service repository $\mathcal{R}$ is defined as follows:*

The states *in $Q^{\mathcal{R}}$ are pairs $\langle \underset{\alpha}{\longmapsfrom} R, q\rangle$, where $\underset{\alpha}{\longmapsfrom} R$ is an activity – i.e. α is an ARN and R is a finite set of requires-points of α – and q is a state of Λ_α.*

The alphabet $A^{\mathcal{R}}$ *is given by pairs $\langle\delta, \iota\rangle$, where $\delta\colon \alpha \to \alpha'$ is a morphism of ARNs and ι is a set of α-actions; thus, $A^{\mathcal{R}}$ is just the alphabet of $\mathcal{A}^\sharp$.*

There exists a transition $\langle \underset{\alpha}{\longmapsfrom} R, q\rangle \xrightarrow{\delta,\iota} \langle \underset{\alpha'}{\longmapsfrom} R', q'\rangle$ *whenever:*

1. *$\langle\alpha, q\rangle \xrightarrow{\delta,\iota} \langle\alpha', q'\rangle$ is a transition of $\mathcal{A}^\sharp$;*
2. *for each requires-point $r \in R$ such that $\xi_r(A_{M_r^+}) \cap \iota \neq \emptyset$ there exists*
   - *a service module $P^r \underset{\alpha^r}{\longleftarrow} R^r$ in $\mathcal{R}$ and*
   - *a polarity-preserving injection $\theta_r : M_r \to M_{P^r}$*

   such that the following colimit can be defined in the category of ARNs

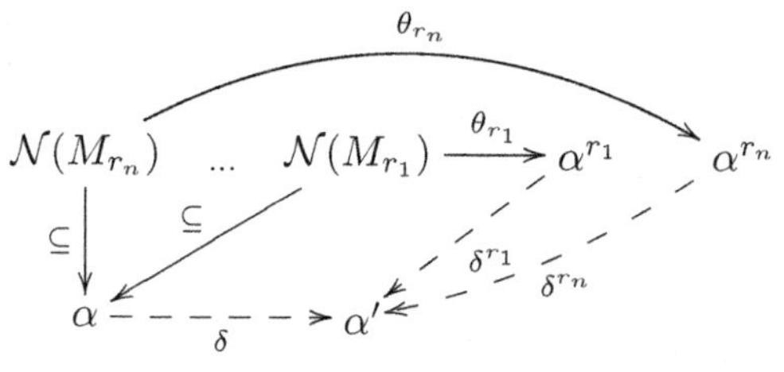

   where $\{r_1, \ldots, r_n\}$ is the biggest subset of R such that $\xi_{r_i}(A_{M_{r_i}^+}) \cap \iota \neq \emptyset$ for all $1 \leq i \leq n$ and $\mathcal{N}(M_{r_i})$ is the atomic ARN that consists of only one point, labelled with the port M_{r_i}, and no hyperedges;
3. *there exists a transition $p' \xrightarrow{\iota'} q'$ of $\Lambda_{\alpha'}$ such that $p'\restriction_\delta = q$, $A_\delta^{-1}(\iota') = \iota$ and, for each requires-point $r \in R$ as above, $p'\restriction_{\delta^r}$ is an initial state of Λ_{α^r}.*

The states *in $I^{\mathcal{R}}$ are those pairs $\langle \underset{\alpha}{\longmapsfrom} R, q\rangle$ for which $q \in I_\alpha$.*
The final-state sets *in $\mathcal{F}^{\mathcal{R}}$ are those sets $\{\langle \underset{\alpha}{\longmapsfrom} R, q\rangle \mid q \in F\}$ for which $F \in \mathcal{F}_\alpha$.*

Note that the definition of the transitions of $\mathcal{R}^\sharp$ integrates both the operational semantics of ARNs given by the functor $\mathcal{A}\colon \mathbb{ARN} \to \mathbb{MA}$ and the logic-programming semantics of service discovery and binding described in [22], albeit in a simplified form, since here we do not take into account the linear temporal sentences that label requires-points. The removal of linear temporal sentences does not limit the applicability of the theory, but rather enables us to give a clearer and more concise presentation of the operational semantics of activities.

Open executions of activities can be defined relative to the automaton $\mathcal{R}^\sharp$ in a similar way to the open executions of ARNs (see Definition 12).

Definition 14 (Open execution of an activity). *An* open execution *of an activity* $\vdash_{\alpha} R$ *with respect to* $\mathcal{R}$ *is an execution of the quasi-automaton* $\mathcal{R}^\sharp$ *that starts from an initial state of* Λ_α*, i.e. a sequence of transitions of* $\mathcal{R}^\sharp$

$$\langle \vdash_{\alpha_0} R_0, q_0 \rangle \xrightarrow{\delta_0, \iota_0} \langle \vdash_{\alpha_1} R_1, q_1 \rangle \xrightarrow{\delta_1, \iota_1} \langle \vdash_{\alpha_2} R_2, q_2 \rangle \xrightarrow{\delta_2, \iota_2} \cdots$$

such that $\alpha_0 = \alpha$*,* $R_0 = R$*, and* $q_0 \in I_\alpha$*. An open execution as above is* successful *if there exists* $i \in \omega$ *such that (a) for all* $j \geq i$*,* $\alpha_j = \alpha_i$*,* $\delta_j = 1_{\alpha_i}$*, and (b)* $\{q_j \mid j \geq i\} \in \mathcal{F}_{\alpha_i}$*.*

To illustrate open executions, let's consider a repository $\mathcal{R}$ formed by the service TravelAgent (depicted in Fig. 4) and the very simple services CurrenciesAgent, AccomodationAgent and FlightsAgent described in Fig. 5. Let's also consider the TravelClient activity of Fig. 3. Observing the automata of Fig. 3 (b) and (c), an execution starts with the activity TravelClient performing one of two actions, *hotels*! or *hotels&Flights*!. Let us assume it is *hotels*! without loss of generality. The prefix of the execution after the transition has the following shape:

$$\langle \vdash_{\mathsf{TravelClient}} \{\mathsf{CC}_1\}, q_0 \rangle \xrightarrow{id, hotels!} \langle \vdash_{\mathsf{TravelClient}} \{\mathsf{CC}_1\}, q_1 \rangle$$

where q_0 and q_1 are the states (q_0, q_0) and (q_1, q_1) of the composed automaton $\Lambda_{TC} \times \Lambda_{CC}$ respectively. After this, the only plausible action in this run is the delivery of the message *hotels* by the communication channel CC. Since $\xi_{\mathsf{TravelClient}}(A_{M^+_{\mathsf{CC}_1}}) \cap \{hotels¡\} = \{hotels¡\}$ this action triggers a reconfiguration of the activity. In our example's repository, $\mathcal{R}$, the only service that can satisfy the requirement CC_1 is TravelAgent. Thus, the action *hotels¡* leads us to the activity TravelClient′ shown in Fig. 6. The prefix of the execution after this last transition is:

$$\cdots \xrightarrow{id, hotels!} \langle \vdash_{\mathsf{TravelClient}} \{\mathsf{CC}_1\}, q_1 \rangle \xrightarrow{\delta, hotels¡} \langle \vdash_{\mathsf{TravelClient}'} \{\mathsf{H}_0, \mathsf{F}_0, \mathsf{CE}_0\}, q_2 \rangle$$

where q_2 is the state $(q_1, q_0, q_1, q_0, q_0)$ of the automaton of TravelClient′. To see that the morphism $\delta : \mathsf{TravelClient} \to \mathsf{TravelClient}'$ exists is straightforward.

A continuation for this execution is obtained by the automaton Λ_{TA}, associated with TravelAgent, publishing the action *getHotels!* and the mandatory delivery *getHotels!* that comes after. This actions trigger a new reconfiguration of the activity on port H_0 of the communication channel C_0; in this case,

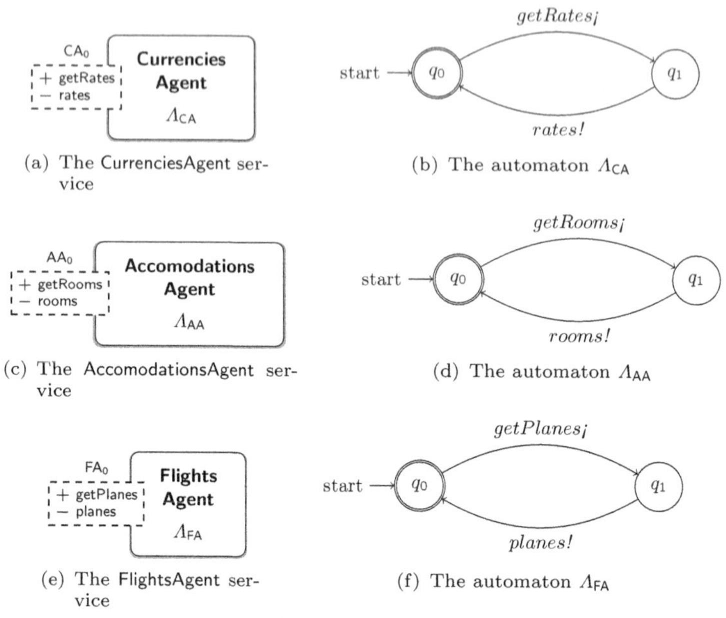

(a) The CurrenciesAgent service

(b) The automaton Λ_{CA}

(c) The AccomodationsAgent service

(d) The automaton Λ_{AA}

(e) The FlightsAgent service

(f) The automaton Λ_{FA}

Fig. 5. Very simple services in $\mathcal{R}$

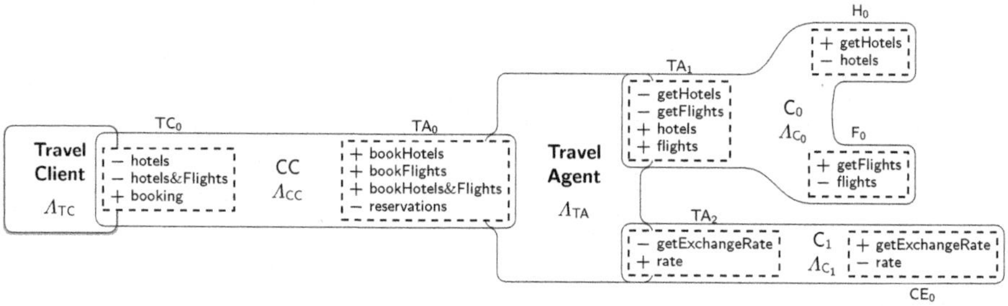

Fig. 6. The TravelClient′ activity

and considering once again our repository $\mathcal{R}$, the result of the reconfiguration should be the *attachment* of the service module AccomodationsAgent.

The following fact allows us to easily generalise Proposition 2 from open executions of ARNs to open executions of activities.

Fact 2. There exists a (trivial) forgetful morphism of Muller automata $\mathcal{R}^\sharp \to \mathcal{A}^\sharp$ that maps every state $\langle \vdash_{\alpha} R, q\rangle$ of $\mathcal{R}^\sharp$ to the state $\langle \alpha, q\rangle$ of $\mathcal{A}^\sharp$.

Proposition 3. *For every (successful) execution*

$$\langle \vdash_{\alpha_0} R_0, q_0\rangle \xrightarrow{\delta_0,\iota_0} \langle \vdash_{\alpha_1} R_1, q_1\rangle \xrightarrow{\delta_1,\iota_1} \langle \vdash_{\alpha_2} R_2, q_2\rangle \xrightarrow{\delta_2,\iota_2} \cdots$$

of the activity quasi-automaton $\mathcal{R}^\sharp$, the infinite sequence

$$q_0 \xrightarrow{\iota_0} q_1\restriction_{\delta_0} \xrightarrow{A_{\delta_0}^{-1}(\iota_1)} q_2\restriction_{\delta_0;\delta_1} \xrightarrow{A_{\delta_0;\delta_1}^{-1}(\iota_2)} \cdots$$

is a (successful) execution of the automaton Λ_{α_0}.

Theorem 1 shows the relation that exists between the traces of an activity with respect to a respository and the automata of each component of the activity. It shows that every (successful) open execution of an activity can be projected to a (successful) execution of each of the automata interveaning. In order to prove that open executions of activities give rise to "local" executions of Λ_{α_i} – for every $i \in \omega$, not only for $i = 0$ – we rely on a consequence of the fact that the functor $\mathcal{A}\colon \mathbb{ARN} \to \mathbb{MA}$ preserves colimits and, in addition, we restrict the automata associated with the underlying ARNs of service modules.

Proposition 4. *The functor $\mathcal{A}\colon \mathbb{ARN} \to \mathbb{MA}$ preserves colimits. In particular, for every transition $\langle \underset{\alpha}{\vdash\!\!-} R, q\rangle \xrightarrow{\delta,\iota} \langle \underset{\alpha'}{\vdash\!\!-} R', q'\rangle$ as in Definition 13, the Muller automaton $\Lambda_{\alpha'}$ is isomorphic with the product*

$$\Lambda_\alpha^\delta \times \prod_{\substack{r\in R\\ \xi_r(A_{M_r^+})\cap\iota\neq\emptyset}} \Lambda_{\alpha^r}^{\delta^r}$$

of the cofree expansions Λ_α^δ and $\Lambda_{\alpha^r}^{\delta^r}$, for $r \in R$ such that $\xi_r(A_{M_r^+}) \cap \iota \neq \emptyset$, of the automata Λ_α and Λ_{α^r} along the alphabet maps A_δ and A_{δ^r}, respectively.[4]

Consequently, a transition $p' \xrightarrow{\iota'} q'$ is defined in the automaton $\Lambda_{\alpha'}$ if and only if $p'\restriction_\delta \xrightarrow{A_\delta^{-1}(\iota')} q'\restriction_\delta$ is a transition of Λ_α and, for each $r \in R$ such that $\xi_r(A_{M_r^+}) \cap \iota \neq \emptyset$, $p'\restriction_{\delta^r} \xrightarrow{A_{\delta^r}^{-1}(\iota')} q'\restriction_{\delta^r}$ is a transition of Λ_{α^r}.

Definition 15 (Idle initial states). *An automaton $\Lambda = \langle Q, 2^A, \Delta, I, \mathcal{F}\rangle$ is said to have* idle initial states *if for every initial state $q \in I$ there exists a transition $(p, \emptyset, q) \in \Delta$ such that p is an initial state too.*

The following result can be proved by induction on $i \in \omega$. The base case results directly from Proposition 3, while the induction step relies on condition 3 of Definition 13 and on Proposition 4.

Theorem 1. *If, for every service module $P \underset{\alpha}{\leftarrow\!\!-} R$ in $\mathcal{R}$, the automaton Λ_α has idle initial states, then for every (successful) execution*

$$\langle \underset{\alpha_0}{\vdash\!\!-} R_0, q_0\rangle \xrightarrow{\delta_0,\iota_0} \langle \underset{\alpha_1}{\vdash\!\!-} R_1, q_1\rangle \xrightarrow{\delta_1,\iota_1} \langle \underset{\alpha_2}{\vdash\!\!-} R_2, q_2\rangle \xrightarrow{\delta_2,\iota_2} \cdots$$

[4] We recall from [22] that the cofree expansion of an automaton $\Lambda = \langle Q, 2^A, \Delta, I, \mathcal{F}\rangle$ along a map $\sigma\colon A \to A'$ is the automaton $\Lambda' = \langle Q, 2^{A'}, \Delta', I, \mathcal{F}\rangle$ for which $(p, \iota', q) \in \Delta'$ if and only if $(p, \sigma^{-1}(X'), q) \in \Delta$.

of $\mathcal{R}^\sharp$ there exists a (successful) execution of Λ_{α_i}, for $i \in \omega$, of the form

$$q'_0 \xrightarrow{\iota'_0} q'_1 \xrightarrow{\iota'_1} \cdots q'_{i-1} \xrightarrow{\iota'_{i-1}} q_i \xrightarrow{\iota_i} q_{i+1}\restriction_{\delta_i} \xrightarrow{A^{-1}_{\delta_i}(\iota_{i+1})} q_{i+2}\restriction_{\delta_i;\delta_{i+1}} \xrightarrow{A^{-1}_{\delta_i;\delta_{i+1}}(\iota_{i+2})} \cdots$$

where, for every $j < i$, $q'_j\restriction_{\delta_j;\cdots;\delta_{i-1}} = q_j$ and $A^{-1}_{\delta_j;\cdots;\delta_{i-1}}(\iota'_j) = \iota_j$.

The reader should notice that all the automata used as examples in this work have *idle initial states* as a consequence of the hidden self loop, labelled with the empty set, that we assumed to exist in every state.

4 Satisfiability of Linear Temporal Logic Formulae

In this section we show how we can use the trace semantics we presented in the previous section to reason about Linear Temporal Logic (LTL for short) [16,19] properties of activities. Next we define linear temporal logic by providing its grammar and semantics in terms of sets of traces.

Definition 16. *Let $\mathcal{V}$ be a set of proposition symbols, then the set of LTL formulae on $\mathcal{V}$, denoted as LTLForm($\mathcal{V}$), is the smallest set S such that:*

- *$\mathcal{V} \subseteq S$, and*
- *if $\phi, \psi \in S$, then $\{\neg\phi, \phi \vee \psi, \mathbf{X}\phi, \phi\mathbf{U}\psi\} \subseteq S$.*

We consider the signature of a repository to be the union of all messages of all the service modules in it. This can give rise to an infinite language over which it is possible to express properties refering to any of the services in the repository, even those that are not yet bound (and might never be). To achieve this we require the alphabets of the service modules in a repository $\mathcal{R}$ to be pairwise disjoint.

Definition 17. *Let $\mathcal{R}$ be a repository and $\underset{\alpha}{\vdash} R$ an activity. We denote with $A_{\mathcal{R},\alpha}$ the set $\left(\bigcup\{A_{\alpha'}\}_{P' \underset{\alpha'}{\leftarrow} R' \in \mathcal{R}}\right) \cup A_\alpha$.*

Defining satisfaction of an LTL formula requires that we first define what is the set of propositions over which we can express the LTL formulae. We consider as the set of propositions all the actions in the signature of the repository or in the activity to which we are providing semantics. Thus, the propositions that hold in a particular state will be the ones that correspond to the actions in the label of the transition that took the system to that state.

In order to define if a run satisfies an LTL formula it is necessary to consider the suffixes of a run, thus let

$$r = \langle \underset{\alpha_0}{\vdash} R_0, q_0\rangle \xrightarrow{\delta_0,\iota_0} \langle \underset{\alpha_1}{\vdash} R_1, q_1\rangle \xrightarrow{\delta_1,\iota_1} \langle \underset{\alpha_2}{\vdash} R_2, q_2\rangle \xrightarrow{\delta_2,\iota_2} \cdots$$

be a succesful open execution of $\underset{\alpha}{\vdash} R$ with respect to a repository $\mathcal{R}$ we denote with r_i the ith suffix of r. That is:

$$r_i = \langle \underset{\alpha_i}{\vdash} R_i, q_i\rangle \xrightarrow{\delta_i,\iota_i} \langle \underset{\alpha_{i+1}}{\vdash} R_{i+1}, q_{i+1}\rangle \xrightarrow{\delta_{i+1},\iota_{i+1}} \langle \underset{\alpha_{i+2}}{\vdash} R_{i+2}, q_{i+2}\rangle \xrightarrow{\delta_{i+2},\iota_{i+2}} \cdots$$

The thoughtful reader may notice that while our formulae are described over the union of the alphabet of the repository $\mathcal{R}$ and the alphabet of the activity $\underset{\alpha}{\longmapsto} R$, the labels ι_i in a run belong to the alphabet A_{α_i}, that is the computed co-limit described in Definition 9. Therefore, we need to *translate* our formula accordingly with the modifications suffered by the activity during the particular run to be able to check if it holds. In order to define how the translation of the formula is carried out we rely on the result of Corollary 1. The following definition provides the required notation to define these translations:

Definition 18. *Let $\mathcal{R}$ be a repository and $\langle \underset{\alpha}{\longmapsto} R, q\rangle \xrightarrow{\delta,\iota} \langle \underset{\alpha'}{\longmapsto} R', q'\rangle$ a transition of $\mathcal{R}^\sharp$ then we define $A_{\hat{\delta}} : A_{\mathcal{R},\alpha} \to A_{\mathcal{R},\alpha'}$ as*

$$A_{\hat{\delta}}(a) = \begin{cases} A_\delta(a) & a \in A_\alpha \\ A_{\delta^{r_i}}(a) & a \in A_{\alpha^{r_i}} \\ a & \textit{otherwise} \end{cases}$$

Definition 19. *Let $\mathcal{R}$ be a repository and let $\underset{\alpha}{\longmapsto} R$ be an activity. Also let $\mathcal{V} = A_{\mathcal{R},\alpha}$, $\phi, \psi \in \mathit{LTLForm}(\mathcal{V})$, $a \in \mathcal{V}$ and $v \subseteq \mathcal{V}$ then:*

- $\langle r, v, \tau\rangle \models \mathbf{true}$,
- $\langle r, v, \tau\rangle \models a$ *iff* $\tau(a) \in v$,
- $\langle r, v, \tau\rangle \models \neg\phi$ *iff* $\langle r, v, \tau\rangle \not\models \phi$,
- $\langle r, v, \tau\rangle \models \phi \vee \psi$ *if* $\langle r, v, \tau\rangle \models \phi$ *or* $\langle r, v, \tau\rangle \models \psi$,
- $\langle r, v, \tau\rangle \models \mathbf{X}\phi$ *iff* $\langle r_1, \iota_0, \tau; A_{\hat{\delta}_0}\rangle \models \phi$, *and*
- $\langle r, v, \tau\rangle \models \phi\mathbf{U}\psi$ *iff there exists* $0 \le i$ *such that* $\langle r_i, \iota_{i-1}, \tau; A_{\hat{\delta}_0}; \ldots; A_{\hat{\delta}_{i-1}}\rangle \models \psi$ *and for all* j, $0 \le j < i$, $\langle r_j, \iota_{j-1}, \tau; A_{\hat{\delta}_0}; \ldots; A_{\hat{\delta}_{j-1}}\rangle \models \phi$ *where* $\iota_{-1} = \emptyset$ *and* $A_{\hat{\delta}_{-1}} = 1_{A_{\mathcal{R},\alpha}}$.

If $\mathcal{V}$ is a set of propositions, $\phi, \psi \in \mathit{LTLForm}(\mathcal{V})$, the rest of the boolean constants and operators are defined as usual as: $\mathbf{false} \equiv \neg\mathbf{true}$, $\phi \wedge \psi \equiv \neg(\neg\phi \vee \neg\psi)$, $\phi \implies \psi \equiv \neg\phi \vee \psi$, etc. We define $\Diamond\phi \equiv \mathbf{trueU}\phi$ and $\Box\phi \equiv \neg(\mathbf{trueU}\neg\phi)$.

Definition 20. *Let $\mathcal{R}$ be a repository and let*

$$r = \langle \underset{\alpha_0}{\longmapsto} R_0, q_0\rangle \xrightarrow{\delta_0,\iota_0} \langle \underset{\alpha_1}{\longmapsto} R_1, q_1\rangle \xrightarrow{\delta_1,\iota_1} \langle \underset{\alpha_2}{\longmapsto} R_2, q_2\rangle \xrightarrow{\delta_2,\iota_2} \cdots$$

be a successful open execution of $\mathcal{R}^\sharp$. Then a formula $\phi \in \mathit{LTLForm}(A_{\mathcal{R},\alpha_0})$ is satisfied by r ($r \models \phi$) if and only if $\langle r, \emptyset, 1_{A_{\mathcal{R},\alpha_0}}\rangle \models \phi$.

Following the previous definitions, checking if an activity $\underset{\alpha}{\longmapsto} R$ satisfies a proposition ϕ under a repository $\mathcal{R}$ is equivalent to checking if every successful open execution of $\underset{\alpha}{\longmapsto} R$ with respect to $\mathcal{R}$ satisfies ϕ.

In the following we will show how the satisfaction relation in Definition 20 can be used to reason about properties of activities. We are particularly interested in asserting properties regarding the future execution of an activity with respect to a repository.

In order to exemplify, let us once again consider the activity TravelClient of Fig. 3(a) and the repository $\mathcal{R}$ formed by the services TravelAgent, CurrenciesAgent, AccomodationAgent, and FlightsAgent described in Figs. 4 and 5. We are then interested in the open successful executions of the quasi-automaton $\mathcal{R}^\sharp$. Two examples of statements we could be interested in are the following properties:

1. Every execution of TravelClient requires the execution of CurrenciesAgent:
   For all successful open executions r of $\mathcal{R}^\sharp$, $r \models \Diamond \left(\bigvee_{a \in A_{M_{\mathsf{CurrenciesAgent}}}} a\right)$.
2. There exists an execution of TravelClient that does not require the execution of FlightsAgent:
   There exists a successful open execution r of $\mathcal{R}^\sharp$, $r \models \Box\left(\neg \bigvee_{a \in A_{M_{\mathsf{FlightsAgent}}}} a\right)$.

The first property is true and it can be checked by observing that in the automaton Λ_{TA} no matter what is the choice for a transition made in the initial state ($bookHotels_¡$, $bookFlights_¡$, or $bookHotels\&Flights_¡$), the transition labelled with action $getExchangeRate!$ belongs to every path that returns to the initial state, that is the only accepting state. Therefore, the reconfiguration of the activity on port CE_0 is enforced in every successful execution.

The second one is also true as it states that there is an execution that does not requires the binding of a flights agent. Observing TravelClient, one can consider the trace in which no order on flights is placed never as the client always choose to order just accommodation.

5 Conclusions and Further Work

The approach that we put forward in this paper combines, in an integrated way, the operational semantics of processes and communication channels, and the dynamic reconfiguration of ARNs. As a result, it provides a full operational semantics of ARNs by means of automata on infinite sequences built from the local semantics of processes, together with the semantics of those ARNs that are selected from a given repository by means of service discovery and binding. Another use for this semantics is in identifying the differences between the nondeterministic behaviour of a component, reflected within the execution of an ARN, and the nondeterminism that arises from the discovery and binding to other ARNs.

In comparison with the logic-programming semantics of services described in [22], this gives us a more refined view of the execution of ARNs; in particular, it provides a notion of execution trace that reflects both internal actions taken by services that are already intervening in the execution of an activity, and dynamic reconfiguration events that result from triggering actions associated with a requires-point of the activity. In addition, by defining the semantics of an activity with respect to an arbitrary but fixed repository, it is also possible to describe and reason about the behaviour of those ARNs whose executions may not lead to ground networks, despite the fact that they are still sound and successful executions of the activity.

The proposed operational semantics allows us to use various forms of temporal logic to express properties concerning the behaviour of ARNs that surpasses those considered before. We showed this by defining a variant of the satisfaction relation for linear temporal logic, and exploiting the fact that reconfiguration actions are observable in the execution traces; thus, it is possible to determine whether or not a given service module of a repository is necessarily used, or may be used, during the execution of an activity formalised as an ARN.

Many directions for further research are still to be explored in order to provide an even more realistic execution environment for ARNs. Among them, in the current formalism, services are bound once and forever. In real-life scenarios services are bound only until they finish their computation (assuming that no error occurs); this does not prevent the activity to require the execution of the same action associated to the same requires-point, triggering a new discovery with a potential different outcome on the choice of the service to be bound. Also, our approach does not consider any possible change on the repository during the execution which leads to a naive notion of distributed execution as simple technical problems can make services temporarily unavailable.

References

1. Ausiello, G., Franciosa, P.G., Frigioni, D.: Directed hypergraphs: problems, algorithmic results, and a novel decremental approach. In: Restivo, A., Ronchi Della Rocca, S., Roversi, L. (eds.) ICTCS 2001. LNCS, vol. 2202, pp. 312–327. Springer, Heidelberg (2001)
2. Barr, M., Wells, C.: Category Theory for Computer Science. Prentice Hall, London (1990)
3. Benatallah, B., Casati, F., Toumani, F.: Representing, analysing and managing Web service protocols. Data Knowl. Eng. **58**(3), 327–357 (2006)
4. Bradbury, J.S., Cordy, J.R., Dingel, J., Wermelinger, M.: A survey of self-management in dynamic software architecture specifications. In: Proceedings of the 1st ACM SIGSOFT Workshop on Self-Managed Systems, WOSS 2004, pp. 28–33. ACM (2004)
5. Brand, D., Zafiropulo, P.: On communicating finite-state machines. J. ACM **30**(2), 323–342 (1983)
6. Bruni, R., Bucchiarone, A., Gnesi, S., Hirsch, D., Lluch Lafuente, A.: Graph-based design and analysis of dynamic software architectures. In: Degano, P., De Nicola, R., Meseguer, J. (eds.) Concurrency, Graphs and Models. LNCS, vol. 5065, pp. 37–56. Springer, Heidelberg (2008)
7. Bruni, R., Bucchiarone, A., Gnesi, S., Melgratti, H.C.: Modelling dynamic software architectures using typed graph grammars. Electr. Notes Theor. Comput. Sci. **213**(1), 39–53 (2008)
8. Bruni, R., Lluch Lafuente, A., Montanari, U., Tuosto, E.: Service oriented architectural design. In: Barthe, G., Fournet, C. (eds.) TGC 2007. LNCS, vol. 4912, pp. 186–203. Springer, Heidelberg (2008)

9. Cambini, R., Gallo, G., Scutellà, M.G.: Flows on hypergraphs. Math. Program. **78**, 195–217 (1997)
10. Clavel, M., Durán, F., Eker, S., Lincoln, P., Martí-Oliet, N., Meseguer, J., Talcott, C. (eds.): All About Maude - A High-Performance Logical Framework. LNCS, vol. 4350, p. 119. Springer, Heidelberg (2007)
11. Fiadeiro, J.L.: Categories for Software Engineering. Springer, Berlin (2005)
12. Fiadeiro, J.L., Lopes, A., Bocchi, L.: An abstract model of service discovery and binding. Formal Aspects Comput. **23**(4), 433–463 (2011)
13. Fiadeiro, J.L., Lopes, A.: An interface theory for service-oriented design. Theor. Comput. Sci. **503**, 1–30 (2013)
14. Gadducci, F.: Graph rewriting for the π-calculus. Math. Struct. Comput. Sci. **17**(3), 407–437 (2007)
15. Jackson, D.: Software Abstractions - Logic, Language, and Analysis. MIT Press, Cambridge (2006)
16. Manna, Z., Pnueli, A.: Temporal Verification of Reactive Systems. Springer, New York (1995)
17. McLane, S.: Categories for Working Mathematician. Graduate Texts in Mathematics. Springer, Berlin (1971)
18. Perrin, D., Pin, J.É.: Infinite Words: Automata, Semigroups, Logic and Games. Pure and Applied Mathematics. Elsevier Science, Amsterdam (2004)
19. Pnueli, A.: The temporal semantics of concurrent programs. Theor. Comput. Sci. **13**(1), 45–60 (1981)
20. Simonot, M., Aponte, V.: A declarative formal approach to dynamic reconfiguration. In: Proceedings of the 1st International Workshop on Open Component Ecosystems, IWOCE 2009, pp. 1–10 (2009)
21. Thomas, W.: Automata on infinite objects. In: van Leeuwen, J. (ed.) Handbook of Theoretical Computer Science, Volume B: Formal Models and Semantics, pp. 133–192. Elsevier, Amsterdam (1990)
22. Ţuţu, I., Fiadeiro, J.L.: Service-oriented logic programming. Logical Methods in Computer Science (in press)

A SOC-Based Formal Specification and Verification of Hybrid Systems

Ning Yu[(✉)] and Martin Wirsing

Programmierung und Softwaretechnik, Institut für informatik, Ludwig-Maximilians-Universität München, Oettingenstrasse 67, 80538 Munich, Germany
{yu,wirsing}@pst.ifi.lmu.de

Abstract. In order to specify hybrid systems in a SOC paradigm, we define Hybrid Doubly Labeled Transition Systems and the hybrid trace of it. Then we extend SRML notations with a set of differential equation-based expressions and hybrid programs and interpret the notations over Hybrid Doubly Labeled Transition Systems. By redefining the dynamic temporal logic dTL, we provide a logic basis for reasoning about the behavior of hybrid transition systems. We illustrate our approach by a case study about the control of a moving train, in which the movement of the train is regulated by ordinary differential equations.

Keywords: Hybrid transition systems · SRML · Differential equations · dTL

1 Introduction

Service-Oriented Computing (SOC) is a computing paradigm that utilizes services as fundamental elements to support rapid, low-cost development of distributed applications in heterogeneous environments [1]. In SOC, a service is defined as an independent and autonomous piece of functionality which can be described, published, discovered and used in a uniform way. Within the development of SOC, complex systems are more and more involved. A typical type of complex systems are the hybrid systems, which arise in embedded control where discrete components are coupled with continuous components. In an abstract point of view, hybrid systems are mixtures of real-time (continuous) dynamics and discrete events [2]. In order to address these two aspects into SOC paradigm, we make our approach by giving a SOC-based formal specification and verification to hybrid systems.

The SOC-based specification of hybrid systems are realized by giving a hybrid extension to the SENSORIA Reference Modeling Language (SRML). SRML is a modeling language that can address the higher levels of abstraction of "business modeling"[3], developed in the project SENSORIA – the IST-FET Integrated Project that develops methodologies and tools such as Web Services [10] for dealing with the challenges arose in Service-Oriented Computing. To make this

M. Codescu et al. (Eds.): WADT 2014, LNCS 9463, pp. 151–169, 2015.
DOI: 10.1007/978-3-319-28114-8_9

approach, we first define: Hybrid Doubly Labeled Transition System (HL^2TSs), which extends the semantic domain of UCTL [11]; hybrid traces of HL^2TSs, which represent traces of the system evolution; and service-oriented Hybrid Doubly Labeled Transition Systems (SO-HL^2TSs), which extends HL^2TSs, as the SRML semantic domain. Then we extend SRML by extending *the language of business role* and *the language of business protocol*. The language of business role is extended by defining formulas and differential equation-based terms for *transition specifications*, and interpreting them over SO-HL^2TSs. The language of business protocol is extended by redefining *hybrid programs* and formulas of the dynamic logic temporal logic dTL [12], which provides modalities for quantifying over traces of hybrid systems, for *behaviour constraints*.

We illustrate our approach though a case study of a *Train-Control system* verification. The Train-Control system abstracts the European Train Control System (ETCS)[19], which is a a signalling, control and train protection system designed to replace the many incompatible safety systems currently used by European railways. In such a system the displacement of the train is continuous on time within the system evolution and is governed by ordinary differential equations. On specifying the system with extended SRML, we verify a safety constraint of it with a set of sequent calculus provided in [12] for verifying hybrid systems.

2 A General Introduction to SRML

In this section we give an overview of SRML composition model and each element of the composition introduced in [5].

SRML is designed for modeling composite services, whose business logic involves a number of interactions among more elementary service components within services and among different services via interfaces. This idea comes from the concepts proposed in Service Component Architectures (SCA)[4]. The basic units of business logic are called *service modules*, which are composed of *service components* and *external interfaces*, and are orchestrated by control and data flows. Service components are computational units central to the composite services. Each service component is modeled by means of the execution pattern that involves a set of interactions and orchestrations related to them. In a service module, external interfaces are used for modeling external parties that either provide services to or require services from this module. Each interface specifies a set of interactions internal to this module and some constraints to which the module expect the external parties to adhere. Service components and external interfaces of the same module are connected to each other through internal *wires* that are used to support and coordinate the interactions among them. Figure 1 shows an example of a service module.

The orchestrations of services components can be seen static, since they are pre-define at design time and do not invoke services of any external party. While the constraints of the external interfaces are dynamically configured at each run time, when modules are discovered and bound to different external parties.

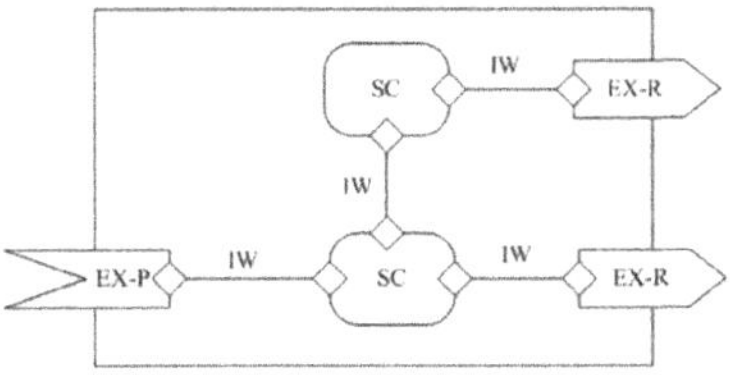

Fig. 1. Service composition

In this paper, we only discuss the way of defining the module, but not the runtime configuration. Next we show in detail the composition of a service module.

Business Role. Service components are specified through business roles, each of which is specified by declaring a set of interactions and the way they are orchestrated. We give the following introduction to each part:

Interactions involve two parties and can be in both directions. They are defined from the point of view of the party in which they are defined. ***Local*** specifies the variables that provide an abstract view of the state of the local component.

Initialization designates a specific initial state.

Transitions model the activities performed by the component. A transition has an optional name and some possible features. These features are classified as follows: (i) A *trigger* is a condition that specifies the occurrence of an event or a state condition; (ii) A *guard* is a condition that identifies the states in which the transition can take place; (iii) Each sentence in *effects* specifies the effects of the transition in the local state.

Business Protocol. External interfaces are specified through business protocols. They declare similar interactions to those in business roles, but from the external parties' point of view. Instead of an orchestration, a business protocol provide a set of properties that the external party is expected to follow.

Behaviour models the behaviors that users can expect form a service. Based on temporal logic [14], they specify which message exchange sequences are supported by the service via a number of ***behaviour constraints***.

Interaction Protocol. Wires that connect service components and external interfaces are specified through interaction protocols, and are labeled by connectors that coordinate the interactions in which the parties are jointly involved. Our work doesn't relate to this part, so it is not introduced in details.

3 Hybrid Extension of SRML

3.1 A Hybrid Extension of SRML Semantic Domain

Since SRML is a control/data flow driven modeling language, the following data signature is adopted as a basic semantic domain:

$$\Omega = \langle D, F \rangle \tag{1}$$

where D is a set of data sorts and F is a $D^* \times D$-indexed family of sets of operations over the sorts. We assume that *time* $\in D$ is a datatype that represents the usual concept of time. And a fixed algebra $\mathcal{U}$ denotes the interpretation of Ω.

SRML is interpreted over Service-oriented Doubly Labeled Transition Systems ($SO\text{-}L^2TSs$), whose structure bases on the UCTL [11] semantic domain – L^2TS. In order to extend SRML over the combination of hybrid systems and transition systems, we define Hybrid L^2TSs (HL^2TS) by extending L^2TSs with a set of continuous functions Σ, and define the trace of a HL^2TS to describe the system evolution. Then we define $SO\text{-}HL^2TS$ over which extended SRML could be interpreted.

Definition 1 (Hybrid Doubly Labeled Transition System). *A hybrid doubly Labelled Transition System* (HL^2TS) is a tuple

$$\langle S, s_0, Act, R, \Sigma, AP, L\rangle$$

where:

- S *is a set of states;*
- $s_0 \in S$ *is the initial state;*
- Act *is a finite set of observable actions;*
- $R \subseteq S \times 2^{Act} \times S$ *is the transition relation. A transition* $(s, \alpha, s') \in R$ *is denoted by* $s \xrightarrow{\alpha} s'$*;*
- Σ *is a set of functions and for every function* $\sigma \in \Sigma, \sigma : [0, r_\sigma] \to S$ *with* $r_\sigma \in \mathbb{R}$ *and* $r_\sigma \geq 0$, σ *is continuous on the interval* $[0, r_\sigma]$*;*
- AP *is a set of atomic propositions;*
- $L : S \to 2^{AP}$ *is a labelling function such that* $L(s)$ *is the subset of all atomic propositions that are true in state* s.

The evolution of a HL^2TS is described by *traces*, which represent sequences of pieces of continuous functions and discrete jumps in the HL^2TS evolution.

Definition 2 (Trace). Let $\langle S, s_0, Act, R, \Sigma, AP, L\rangle$ be a *HL^2TS* then*:*

- *For every* $\sigma \in \Sigma$, $\sigma[0, r_\sigma]$ *denotes the trace of infinitely many states* $\sigma(0), \ldots,$ $\sigma(r_\sigma)$ *along* σ *over the interval* $[0, r_\sigma]$*;*
- ρ *is a hybrid trace from* s_0 *if* $\rho = (s_0 \xrightarrow{\alpha_0} \sigma_1, \sigma_1[0, r_{\sigma_1}], \sigma_1(r_{\sigma_1}) \xrightarrow{\alpha_1} \sigma_2(0),$ $\sigma_2[0, r_{\sigma_2}], \ldots)$ *where* $(s_0, \alpha_0, \sigma_1(0)) \in R$ *and* $\sigma_i(r_{\sigma_i}), \alpha_i, \sigma_{i+1}(0)) \in R$ *with* $i \in \mathbb{N}$*;*
- *A position* of ρ is a pair *(i, ζ)* with $i \in \mathbb{N}$ and ζ in the interval $[0, r_{\sigma_i}]$*;* the state or transition of ρ at (i, ζ) *is* $\sigma_i(\zeta)$. Positions of ρ are ordered lexicographically by $(i, \zeta) \prec (j, \xi)$ iff either $i < j$, or $i = j$ and $\zeta < \xi$;
- *A trace* ρ *starting from the initial state* s_0 *terminates if it is a finite sequence* $(\rho = (s_0 \xrightarrow{\alpha} \sigma_1, \sigma_1[0, r_{\sigma_1}], \sigma_1(r_{\sigma_1}) \xrightarrow{\alpha_1} \sigma_2(0), \sigma_2[0, r_{\sigma_2}], \ldots, \sigma_n[0, r_{\sigma_n}])$*, and the first state of the trace* s_0 *is denoted by* $first\rho$*, the last state of the trace* $\sigma_n(r_{\sigma_n})$ *is denoted by* $last\rho$*;*

– *The concatenation of traces* $\rho_1 = (s_1 \xrightarrow{\alpha_0} \sigma_1(0), \sigma_1[0, r_{\sigma_1}], \ldots)$ *and* $\rho_2 = (s_2 \xrightarrow{\alpha'_0} \varsigma_1(0), \varsigma_1[0, r_{\varsigma_1}], \ldots)$*, denoted by* $\rho_1 \circ \rho_2$*, is defined as follows:*

$$\rho_1 \circ \rho_2 = \begin{cases} (s_0 \xrightarrow{\alpha_0} \sigma_1(0), \ldots, & \text{if } \rho_1 \text{ terminates at } \sigma_n(r_{\sigma_n}) \\ \quad \sigma_n[0, r_{\sigma_n}], s'_0 \xrightarrow{\alpha'_0} \varsigma_1(0), \ldots) & \quad \text{and } \sigma_n(r_{\sigma_n}) = s'_0 \\ \rho_1 & \text{if } \rho_1 \text{ does not terminate} \\ \text{not defined} & \text{otherwise} \end{cases}$$

SO-HL²TSs extends HL²TSs and Service-Oriented Transition Systems (SO-TSs). SO-TSs are defined in [6]. By combining these two types of transition systems, we get the semantic domain for SRML extension.

Definition 3 (Service-Oriented HL²TSs). *The Service-Oriented HL² TS (denoted SO-HL² TS) that abstracts a SO-TS* $\langle S, \rightarrow, s_0, G\rangle$ *is the tuple*

$$\langle S, s_0, Act, R, \Sigma, AP, L, TIME, \Pi\rangle$$

where:

- $\langle S, s_0, Act, R, \Sigma, AP, L\rangle$ *is the corresponding HL² TS;*
- $Act = \{e! : e \in E\} \cup \{e¡ : e \in E\} \cup \{e? : e \in E\} \cup \{e¿ : e \in E\}$;
- $R \subseteq S \times 2^{Act} S$ *is such that:*
  - $s \xrightarrow{\alpha} s'$ *iff* $(s, \alpha, s' \in R$ *for some* $\alpha \in Act^2$;
  - *For every* $(s), \alpha, s') \in R$:

$$\begin{aligned}\alpha =& \{e! : e \in PUB^{s \xrightarrow{\alpha} s'} \cup \{e¡ : e \in ADLV^{s \xrightarrow{\alpha} s'}\} \cup \\ & \{e? : e \in EXC^{s \xrightarrow{\alpha} s'}\} \cup \{e¿ : e \in DSC^{s \xrightarrow{\alpha} s'}\}\end{aligned}$$

- $AP = \{e! : e \in E\} \cup \{e¡ : e \in E\} \cup \{e? : e \in E\} \cup$ $\{e¿ : e \in E\} \cup \{a.pledge : a \in 2WAY\}$;
- $L : S \rightarrow 2^{AP}$ *is such that:*

$$\begin{aligned}L(s) =& \{e! : e \in HST!^s\} \cup \{e¡ : e \in HST¡^s\} \cup \\ & \{e? : e \in HST?^s\} \cup \{e¿ : e \in HST¿^s\} \\ & \cup PLG^s\end{aligned}$$

with $s \in S$;
- $TIME$ *assigns to each state* $s \in S$ *the instant* $TIME^s$;
- Π *assigns to each state* $s \in S$ *the parameter interpretation* Π^s.

In Definition 3, E is the set of all events of a configuration defined in [6]. a is an interaction and $a.pledge$ is the pledge that is associated with that interaction in the configuration. $2WAY$ is the set of interactions that take place in both directions in the configuration. $HST!, HST¡, HST?$ and $HST¿$ are subsets of events in a computation state, PLG is the set of pledges that holds in the computation state and $TIME \in time_{\mathcal{U}}$. $PUB, ADLV, EXC$ and DSC are subsets of events in a computation step, where computation state and computation step are used to describe the computation of a configuration and they are also defined in [6].

In the rest of this section, we present the semantics of SRML extension interpreted over $SO\text{-}HL^2TSs$ defined in Sect. 2.2. The SRML extension consists of two parts: the extension of business role and the extension of business protocol. The latter includes an extension of dTL formulas which is used to specify and verify *behaviours* specified by the new extended *behaviour constraint* in business protocol.

In order to define the SRML extension, throughout the remaining of this section we consider:

- $sig = \langle NAME, PARAM \rangle$ (defined in [6]) to be an interaction signature where Act is the set of actions associated with sig;
- VAR (defined in [6]) to be an attribute declaration.
- $\Xi = \langle N, W, PLL, \Psi, 2WAY, 1WAY \rangle$ (defined in [6]) to be a configuration;
- II (defined in [6]) to be an interaction interpretation for sig over $2WAY \cup 1WAY$ local to some node $n \in N$;
- $tr = \langle S, s_0, Act, R, \Sigma, AP, L, TIME, \Pi \rangle$ to be a SO-HL^{2}TS for Ξ;
- Δ (defined in [6]) to be an attribute interpretation for VAR over m;
- $m = \langle N, W, C, client, spec, prov \rangle$ (defined in [6]) to be a service module.

3.2 Business Role Extension

Business role is defined over sig and VAR. We extend it by introducing new formulas and predicates into *transitions* (see Fig. 3). These formulas and predicates are defined based on a set of terms.

State terms denote the values of the variables and parameters of events in states. They are interpreted over states.

Definition 4 (State Terms). *The D – indexed family of sets $STERM$ of state terms is defined as follows:*

- *If $c \in F_d$, then $c \in STERM_d$ for every $d \in D$;*
- *If $f \in F_{<d_1,\ldots,d_{n+1}>}$ and $\overrightarrow{p} \in STERM_{<d_1,\ldots,d_n>}$, then $f(\overrightarrow{p}) \in STERM_{d_{n+1}}$ for every $d_1, \ldots, d_n, d_{n+1} \in D$;*
- *If $a \in NAME$ and $param \in PARAM(a)_d$, then $a.param \in STERM_d$ for every $d \in D$;*
- *If $t \in VAR_{time}$, then $t \in STERM_{time}$;*
- *If $v \in VAR_d$, then $v \in STERM_d$ for every $d \in D$ and $d \neq time$.*

For example in Fig. 2, in the **guard** of **transition** *negotiation*, terms "m", "*currentDis*" and "ST" are state terms.

Definition 5 (Interpretation of State Terms). *The interpretation of a state term $T \in STERM$ in a state $s \in S$, written $[\![T]\!]_s$, is defined as follows, where view is the function that defines how the parameter is observed:*

```
BUSINESS ROLE Train is
  INTERACTIONS
    rcv getDisplacement
       🔔 dis:displacement

    snd reportPosition

       🔔 p:displacement

    r&s MAControl

    r&s moveOn

  ORCHESTRATION
  local a:acceleraction,

          currentDis:displacement,

          End:[0,1], v:speed

  initialisation

       End=0, v=normalSpeed, currentDis=0, TIME:time

   transition pointPosition
     | trigger getDisplacement🔔
     | guard m-currentDis≥ST
     | effect1 reportPosition🔔
     |   ∧ reportPosition🔔.p=getDisplacement🔔.dis
     |   ∧ currentDis=getDisplacement🔔.dis
    transition negotiation
     | trigger getDisplacement🔔
     | guard m-currentDis<ST
     | effect1 TIME=t'
     | effect2 C_time=V
     |   ∧     l_time=1
    transition correction
     | triggeredBy MAControl🔔
     | guardedBy m-currentDis<ST
     | effect1 MAControl✉
     |   ∧     TIME=t'
     | effect2 C_time=v
     |   ∧     V_time=-b
   transition pointPosition
     | triggeredBy moveOn🔔
     | effects1 End=1
```

Fig. 2. Business role: train

- $[\![c]\!]_s = c_{\mathcal{U}}$
- $[\![f(T_1, \ldots, T_n)]\!]_s = f_{\mathcal{U}}([\![T_1]\!]_s, \ldots, [\![T_n]\!]_s)$

- $[\![a.param]\!]_s = view(II(a).param'^{\Pi^s})$
- $[\![t]\!]_s = TIME^s$
- $[\![v]\!]_s = v^{\Delta(s)}$

State predicates are defined based on state terms, and specify the properties of states. The satisfaction of state predicates is defined for states.

Definition 6 (State Predicates). *The state predicates* SP *is defined as follows:*

$$\chi ::= T_1 = (>, <, \neq) T_2 \mid \chi \wedge \chi' \mid \neg \chi \mid \chi \rightarrow \chi'$$

with $T_1, T_2 \in TERM_d$ *for some* $d \in D$.

For example in Fig. 2, **guard** $m - currentDIS \geq ST$ of **transition** *negotiation* is a state predicate.

Definition 7 (Satisfaction of State Predicates). *The satisfaction relation of state predicates is defined for every state* $s \in S$ *follows:*

- $s \models T_1 = (>, <, \neq) T_2$ *iff* $[\![T_1]\!]_s = (>, <, \neq)[\![T_2]\!]_s$
- $s \models \chi \wedge \chi'$ *iff* $s \models \chi$ *and* $s \models \chi'$
- $s \models \neg \chi$ *iff not* $s \models \chi$
- $s \models \chi \rightarrow \chi'$ *iff* $s \models \chi \rightarrow s \models \chi'$

Effect terms denote the values of the variables and parameters of events in transitions, so terms denoting variable values in the source state$(v, time)$ and in the target state $(v', time')$ within a transition are included. Effect terms are interpreted over transitions.

Definition 8 (Effect Terms). *The D-indexed family of sets ETERM of effect terms is defined inductively as follows:*

- *The effect terms* c, $f(\overrightarrow{p})$ *and* $a.param$ *are defined the same way as state terms;*
- *If* $t \in VAR_{time}$ *then* $t, t' \in ETERM_{time}$*;*
- *If* $v \in VAR_d$, *then* $v, v' \in ETERM_d$ *for every* $d \in D$ *and* $d \neq time$.

For example in Fig. 2, in **effect1** of **transition** *pointPosition*, terms "C_0", "*currentDis*", "*reportPosition*🔔*.p*" and "*getDisplacement*🔔*.dis*" are effect terms.

Definition 9 (Interpretation of Effect Terms). *The interpretation of an effect term* $T \in ETERM$ *over a transition* $s \xrightarrow{\alpha} s'$ *written* $[\![T]\!]_{s \xrightarrow{\alpha} s'}$ *is defined as follows, where:* $II(param) = \langle param', view \rangle$*:*

- $[\![c]\!]_{s \xrightarrow{\alpha} s'} = c_{\mathcal{U}}$
- $[\![f(T_1, \ldots, T_n)]\!]_{s \xrightarrow{\alpha} s'} = f_{\mathcal{U}}([\![T_1]\!]_{s \xrightarrow{\alpha} s'}, \ldots, [\![T_n]\!]_{s \xrightarrow{\alpha} s'})$
- $[\![a.param]\!]_{s \xrightarrow{\alpha} s'} = view(II(a).param'^{\Pi^{\sigma'(0)}})$
- $[\![v]\!]_{s \xrightarrow{\alpha} s'} = v^{\Delta(s)}$
- $[\![v']\!]_{s \xrightarrow{\alpha} s'} = v^{\Delta(s')}$

- $[\![t]\!]_{s \xrightarrow{\alpha} s'} = TIME^{\sigma(s)}$
- $[\![t']\!]_{s \xrightarrow{\alpha} s'} = TIME^{s'}$

Effect formulas are defined based on effect terms, and specify the effects of state transitions. The satisfaction relation of effect formulas is defined for transitions.

Definition 10 (Effect Formulas). *The Effects Formulas* EF *is defined as follows:*

- $\chi ::= true \mid T_1 = T_2 \mid ini \mid \chi \wedge \chi' \mid \neg\chi$

where $T_1, T_2 \in ETERM_d$ *for some* $d \in D$, *and* $ini \in En^{INI}$.

For example in Fig. 2, $c_0 = getDisplacement$🔔$.dis$ in **effect1** of **transition** *pointPosition* is an effect formula.

Definition 11 (Satisfaction of Effect Formulas). *The satisfaction relation of effect formulas* EF *is defined for every transition* $s \xrightarrow{\alpha} s'$ *as follows:*

- $s \xrightarrow{\alpha} s' \models true$
- $s \xrightarrow{\alpha} s' \models T_1 = T_2$ *iff* $[\![T_1]\!]_{s \xrightarrow{\alpha} s'} = [\![T_2]\!]_{s \xrightarrow{\alpha} s'}$
- $s \xrightarrow{\alpha} s' \models ini$ *iff* $II(ini) \in PUB^{s \xrightarrow{\alpha} s'}$
- $s \xrightarrow{\alpha} s' \models \chi \wedge \chi'$ *iff* $s \xrightarrow{\alpha} s' \models \chi$ *and* $s \xrightarrow{\alpha} s' \models \chi'$
- $s \xrightarrow{\alpha} s' \models \neg\chi$ *iff not* $s \xrightarrow{\alpha} s' \models \chi$

Extended effect terms denote the values of variables along a trace of state $\sigma[0, r_\sigma]$; They extend effect terms by introducing the term v_{time}, which is used to denote the time derivative of variable v at any time point $TIME^{\sigma(\zeta)}$. Where $\sigma \in \Sigma$ and $\zeta \in [0, r_\sigma]$. They are interpreted along traces.

Definition 12 (Extended Effect Terms). *The D-indexed family E-ETERM of sets of extended effect terms is defined inductively as follows:*

- *The extended effect terms* $c, f(\overrightarrow{p}), t, t', v$ *and* v' *are defined the same way as effect terms;*
- *If* $v \in VAR_d$, *then* $v_{time} \in$ *E-ETERM*$_{d'}$.

For example in Fig. 2, in **effect2** of **transition** *correction*, terms "C_{time}", "V", "V_{time}" and "$-b$" are extended effect terms.

Definition 13 (Interpretation of Extended Effect Terms). *An extended semantics of an effect term* $T \in$ *E-ETERM is interpreted along a trace* $\sigma(0, r_\sigma)$, *written* $[\![T]\!]_{\sigma(0,r_\sigma)}$ *is defined as follows:*

- $[\![c]\!]_{\sigma(0,r_\sigma)} = c_{\mathcal{U}}^{\sigma(0,r_\sigma)}$
- $[\![f(T_1, \ldots, T_n)]\!]_{\sigma(0,r_\sigma)} = f_{\mathcal{U}}([\![T_1]\!]_{\sigma(0,r_\sigma)}, \ldots, [\![T_n]\!]_{\sigma(0,r_\sigma)})$
- $[\![v_{time}]\!]_{\sigma(0,r_\sigma)} = v_{time}^{\Delta(\sigma(0,r_{\sigma'}))}$
- $[\![v]\!]_{\sigma(0,r_\sigma)} = v^{\Delta(\sigma(0,r_{\sigma'}))}$

- $[\![t]\!]_{\sigma(0,r_\sigma)} = TIME^{\sigma(0,r_{\sigma'})}$

Extended effect formulas are defined based on extended effect terms, and specify the first order differential equations about certain variables and time (in Fig. 3 for example, in transition *negotiation*, $C_{time} = v_0$ is a differential equation about the displacement of a train C and time, where C is a globally defined variable). The satisfaction relation of extended effect formulas is defined for traces of states.

Definition 14 (Extended Effect Formulas). *The Extend Effects Formulas E-EF is defined as follows:*

- $\chi ::= true \mid v_{time} = T \mid \chi \wedge \chi' \mid \neg\chi$

where $v_{time}, T \in$ *E-ETERM.*

For example in Fig. 2, $C_{time} = V$ in **effect2** of **transition** *correction* is an extended effect formula.

Definition 15 (Satisfaction for Extended Effect Formulas). *The satisfaction relation for the extended effect formulas E-EF is defined for every trace* $\sigma[0, r_\sigma]$ *as follows:*

- $\sigma[0, r_\sigma] \models true$
- $\sigma[0, r_\sigma] \models v_{time} = T$ *iff* v *is continuous over* $TIME^{\sigma[0,r_\sigma]}$ *and has a time derivative of value* $[\![T]\!]_{\sigma(\zeta)}$ *at each state* $\sigma(\zeta)$ *with* $\zeta \in (0, r_\sigma)$;
- $\sigma[0, r_\sigma] \models \chi \wedge \chi'$ *iff* $\sigma[0, r_\sigma] \models \chi$ *and* $\sigma[0, r_\sigma] \models \chi'$
- $\sigma[0, r_\sigma] \models \neg\chi$ *iff not* $\sigma[0, r_\sigma] \models \chi$

Using state predicates, effect formulas and extended effect formulas, we can specify a transition of a business role component in a SO-HL^{2}TS.

Definition 16 (Transition Specification). *A transition specification is a triple*

$$\langle trigger, guard, effect1, effect2\rangle$$

where $trigger \in Act$, $guard \in SP$, $effect1 \in EF$ *and* $effect2 \in$ *E-EF.*

The satisfaction relation of transitions is defined for SO-HL^{2}TSs.

Definition 17 (Transition Satisfaction). *The SO-HL^{2}TS* m *satisfies a transition specification*

$$r = \langle trigger, guard, effects1, effect2\rangle$$

written $m \models r$, *iff for every transition* $s \xrightarrow{\alpha} \sigma(0)$ *the following property hold: If* $II(trigger) \in \alpha$, $s \models guard$, *then*

$$s \xrightarrow{\alpha} \sigma(0) \models effect1 \text{ and } \sigma[0, r_\sigma] \models effect2.$$

3.3 Business Protocol Extensions

As introduced in Sect. 2.1, the behaviors of business protocols are specified through a set of behavior constraints. We introduce a new behavior constraint (see Fig. 4: **always** $l \leq L \to C < m$) by defining *hybrid programs* and *dTL formulas*. The behavior constraint captures common requirements along all the traces of a system run.

```
BUSINESS PROTOCOL RadioBlockCentre is
  INTERACTIONS
    rcv getPosition
        p:displacement
    s&r systemBusy
    s&r newMA

  BEHAVIOUR
    getPosition? enables systemBusy✉?
    always l ≤ L → C < m
```

Fig. 3. Business protocol: RadioBlockCentre

Hybrid programs [8] generalize real-time programs [9] to hybrid change and are used to describe the behaviour of hybrid systems. They provide uniform discrete jumps and continuous evolutions along differential equations. In [12], hybrid programs are defined over a set variables and terms and used to specify dTL formulas. In this paper, we define hybrid programs with transitions specified in last section, and interpret them along hybrid traces.

Definition 18 (Redefined Hybrid Programs). *The set of hybrid programs HP of a* SO-HL^2 *TSs* m *is inductively defined as follows:*

- *If a transition* $r = \langle trigger, guard, effect1, effect2\rangle$ *over* m *has the satisfaction relation* $m \models r$*, then* $r \in HP$*;*
- *If* $\beta \in HP$*, then* $first\beta \in HP$*;*
- *If* $\beta, \gamma \in HP$*, then* $(\beta \cup \gamma) \in HP$*;*
- *If* $\beta, \gamma \in HP$*, then* $(\beta; \gamma) \in HP$*;*
- *If* $\beta \in HP$*, then* $(\beta^*) \in HP$*;*

Definition 19 (Trace Semantics of Redefined Hybrid Program). *The trace semantics of a redefined hybrid program* β*, written* $[\![\beta]\!]$*, is defined as follows:*

- $[\![trigger, guard, effect1, effect2]\!] = \{s \xrightarrow{\alpha} \sigma(0), \sigma[0, r_\sigma] : trigger \in \alpha, s \models guard, s \xrightarrow{\alpha} \sigma(0) \models effect1$ *and* $\sigma[0, r_\sigma] \models effect2\}$*;*

- $[\![first\beta]\!] = \{s \xrightarrow{\alpha} \sigma(0), \sigma[0,0] : (s \xrightarrow{\alpha} \sigma(0), \sigma[0, r_\sigma]) \in [\![\beta]\!]\}$
- $[\![\beta \cup \gamma]\!] = [\![\beta]\!] \cup [\![\gamma]\!]$;
- $[\![\beta;\gamma]\!] = \{\sigma \circ \varsigma : \sigma \in [\![\beta]\!], \varsigma \in [\![\gamma]\!]$ *when* $\sigma \circ \varsigma$ *is defined*$\}$;
- $[\![\beta^*]\!] = \bigcup_{n \in \mathbb{N}[\![\beta^n]\!]}$, *where* $\beta^{n+1} = (\beta^n;\beta)$ *for* $n \geq 1$.

Given a service module $m = \langle N, W, C, client, spec, prov\rangle$, the function $hp : m \rightarrow HP$ maps SRML specifications into hybrid programs. hp is constructed similar to the method provided in [6] (for details see [7]). For example, the hybrid program of the business role component *Train* in module *Train-Control* (see Fig. 2) is: $hp(Train) = [pointPosition^*] \cup [pointPosition^*; negotiation]$ $\cup [pointPosition^*; negotiation; correction]$.

Based on the definition of hybrid programs, we redefine dTL formulas as follows:

Definition 20 (Redefined dTL Formulas). *The sets* Fml *of dTL state formulas and* Fml_T *of dTL trace formulas are inductively defined as the smallest set such that* ($\phi \in Fml$ *and* $\pi \in Fml_T$):

$$\phi ::= true \mid sp \mid \neg\phi \mid \phi \wedge \phi' \mid \phi \vee \phi' \mid \phi \rightarrow \phi' \mid \forall t\phi \mid \exists t\phi \mid [\beta]\pi \mid \langle\beta\rangle\pi$$
$$\pi ::= \phi \mid \Box\phi \mid \Diamond\phi$$

with $sp \in SP$, $t \in VAR_{time}$ and $\beta \in HP$.

Formulas without $\Box$ and $\Diamond$ are called *non-temporal* d$\mathcal{L}$ *formulas* [8]. Unlike in UCTL, state formulas are true on a trace if they hold for the last state of that trace but not for the first. Thus, $[\beta]\phi$ expresses that ϕ is true at the end of each trace of β. In contrast, $[\beta]\Box\phi$ expresses that ϕ is true all along all states of every trace of β. According to the valuation of dTL formulas defined in [12], we define the semantics of dTL formulas as follows:

Definition 21 (Satisfaction of dTL Formulas). Let $\langle S, s_0, Act, R, \Sigma, AP, L, TIME, \Pi\rangle$ *be a SO-*HL^2 *TS. The satisfaction relation for dTL state formulas on each state* $s \in S$ *is defined as follows, where* $s[t \mapsto \tilde{t}]$ *denotes the modification that agrees with state s on all variables except for variable* $t \in VAR_{time}$:

- $s \models true$;
- $s \models sp$ iff $sp \in L(s)$
- $s \models \neg\phi$ *iff not* $s \models \phi$;
- $s \models \phi \wedge \phi'$ *iff* $s \models \phi$ *and* $s \models \phi'$;
- $s \models \phi \vee \phi'$ *iff* $s \models \phi$ *or* $s \models \phi'$;
- $s \models \phi \rightarrow \phi'$ *iff* $s \models \phi \rightarrow s \models \phi'$;
- $s \models \forall t\phi$ *iff* $s[t \mapsto \tilde{t}] \models \phi$ *for all* $\tilde{t} \in VAR_{time}$;
- $s \models \exists t\phi$ *iff* $s[t \mapsto \tilde{t}] \models \phi$ *for some* $\tilde{t} \in VAR_{time}$;
- $s \models [\beta]\pi$ *iff for each trace* $\rho \in [\![\beta]\!]$ *with* $first\rho = s$, *if the satisfaction relation between* ρ *and* π *is defined then* $\rho \models \pi$;
- $s \models \langle\beta\rangle\pi$ *iff there is a trace* $\rho \in [\![\beta]\!]$ *with* $first\rho = s$, *if the satisfaction relation between* ρ *and* π *is defined then* $\rho \models \pi$.

The satisfaction relation for dTL trace formulas with respect to trace $\rho = (s \xrightarrow{\alpha} \sigma_1, \sigma_1[0, r_{\sigma_1}], \sigma_1(r_{\sigma_1}) \xrightarrow{\alpha_1} \sigma_2(0), \sigma_2[0, r_{\sigma_2}], \ldots)$ *is defined as follows where* ϕ *is a state formula and* Λ *denotes the failure of a system run:*

- $\rho \models \phi$ *iff* ρ *terminates and* $last\rho \models \phi$*, whereas the satisfaction relation between* ρ *and* ϕ *is not defined if* ρ *does not terminates;*
- $\rho \models \Box\phi$ *iff* $\sigma_i(\zeta) \models \phi$ *for all positions* (i, ζ) *of* ρ *with* $\sigma_i(\zeta) \neq \Lambda$*;*
- $\rho \models \Diamond\phi$ *iff* $\sigma_i(\zeta) \models \phi$ *for some positions* (i, ζ) *of* ρ *with* $\sigma_i(\zeta) \neq \Lambda$*;*

In the end we can define the new behaviour constraint ***always*** *s* for extending business protocol:

Definition 22 (Hybrid Behaviour Constraint). *For any service module* m *with the corresponding sets* C *and* $WW \in W$*:*

- *"**always** sp" stands for*

$$[hp(C, WW)]\Box sp$$

(sp is true in each state along every trace of hybrid program hp(C, WW) starting from the initial state, where $sp \in SP$*).*

For example in Fig. 3, the **behaviour "alwalys** $l < L \to C < m$" stands for $[hp(Train)]\Box(l \leq L \to C < m)$.

4 Case Study: The Verification of Train-Control System

The model of Train-Control System is inspired by the European Train Control System(ETCS). As shown in Fig. 4, the system is composed by three components: Train, Radio Block Centers (RBC) and Balise which is melded with the railway. RBC grant or deny movement authorities (MA) to individual train by wireless communication. A train can not exceed the current MA (say m) in order to guarantee safety driving. The balise reports to the train its current position periodically, so the train knows how far it still is from the end of MA. Before entering negotiation at some point ST (in the "*far*" region), the train has sufficient distance to MA and can regulate its speed freely. When the train enters the region "*neg*", it sends a request to the RBC to apply for the MA extension and proceed with a constant speed v_0. If the train receives negative response from the RBC, it enters the "*cor*" region and proceed with acceleration $-b$. With the restriction of the scenario above, we have the hybrid program $hp(Train - ControlSystem(Train)) = [pointPosition^*] \cup [pointPosition^*; negotiation] \cup [pointPosition^*; negotiation; correction]$.

Figure 5 shows the SRML module *Train-Control*. Each element of the module is described as follows:

- business role: TR – a component that coordinates the movement process of the train, of type *Train*;

- business protocol: RBC– the external interface of the module which provides service to the external parties for knowing the current position of the train and issuing movement authority, of type *RadioBlockCentre*;
- business protocol: BA – the external interface of the module which requires service from the external parties for getting the current positioning signal, of type *Balise*;
- interaction protocol: RT, TB – two internal wires that make the partner relationship between RBC and TR, TR and BA explicitly.

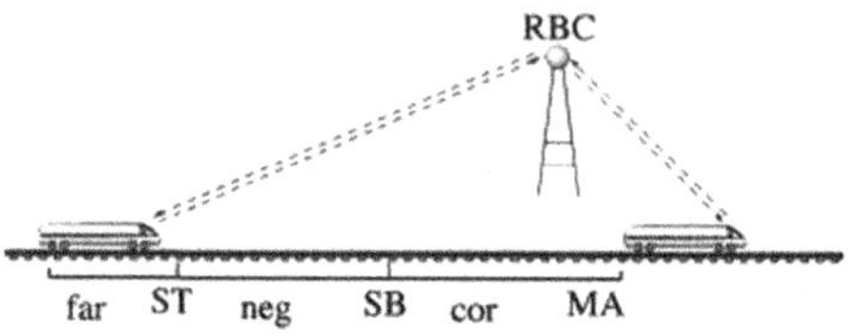

Fig. 4. ETCS train coordination

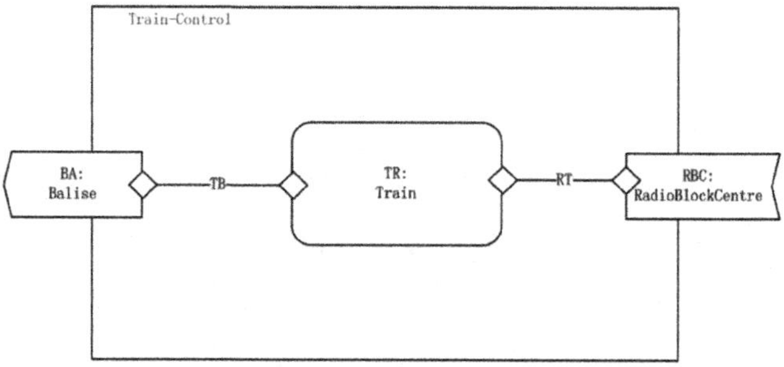

Fig. 5. Train-Control module

The whole SRML specification can be found in Appendix 2. In business role *Train*, C : *displacement* is a global variable of the train displacement which is continuous in time. Differential equation $C_{time} = v_0$ describes the movement of the train in region "(neg)"; and differential equation set $C_{time} = V, V_{time} = -b$ describes the movement of the train in region "cor".

Next we show the verification of the behaviour, **always** $l < L \to C < m$, specified in business protocol *RadioBlockCentre*. This behaviour expresses that,under a initial condition ϕ for parameters, a train will always remain within its MA m, as long as the accumulated RBC negotiation latency l is at most L. We assume that every transition $\sigma(t_\sigma) \xrightarrow{\alpha} \sigma'(0)$ takes so short time that we could approximately have $C^{\Delta(\sigma(r_\sigma))} = C^{\Delta(\sigma'(0))}, l^{\Delta(\sigma(r_\sigma))} = l^{\Delta(\sigma'(0))}$.

So we have $\psi \to [run(\text{Train-Control System})]\Box(l \leq L \to C < m)$, where ψ is the set of initial propositions and $run(\text{Train-Control System}) = [pointPosition^*]$

$\cup$ $[pointPosition^*; negotiation]$ $\cup$ $[pointPosition^*; negotiation; correction]$. According to the scenario, the train is in region "*far*" along trace $pointPosition^*$, we always have $C < m$. Thus the proof of this part can be omitted. In the end we have the following formula:

$$\begin{aligned} \psi &\rightarrow [negotiation; correction]\Box\phi \qquad (2)\\ \phi &\equiv l \leq L \rightarrow C < m \\ \psi &\equiv C_0 < m \wedge v_0 > 0 \wedge l_0 = 0 \wedge L \geq 0 \\ negotiation &\equiv getDisplacement\text{🔔}, m - currentdis < ST, true, \\ &\quad C_{time} = v_0 \wedge l_{time} = 1 \\ correction &\equiv MAControl\text{🔔}, m - currentdis < ST, MAControl\boxtimes, \\ &\quad C_{time} = V \wedge V_{time} = -b \end{aligned}$$

where C_0 is the initial displacement and l_0 is the initial negotiation latency. As shown in Fig. 2, the train first negotiate with RBC while keeping a constant speed V and the movement is controlled by equation ($C_{time} = V$) in transition *negotiation*. Then in transition *correction* the train brakes with acceleration $-b$ and the movement is controlled by equations ($C_{time} = V \wedge V_{time} = -b$).

We use the rule schema of the dTL calculus provided in [7] for our proof. The rule schema can be find in Appendix 1 and $\{\!\!\{\cdot\}\!\!\}$ brackets are used instead of modalities to visually identify the update prefix. We omit all the events, variables and parameters that don't appear in the state predicate $l < L \rightarrow C < m$.

The dTL proof of the constraint in (2) splits into two cases as follows:

$$\text{P3}\ \dfrac{\text{T1}\ \dfrac{\dfrac{\cdots}{\psi \vdash [negotiation]\Box\phi} \qquad \dfrac{\cdots}{\psi \vdash [negotiation][correction]\Box\phi}}{\psi \vdash [negotiation; correction]\Box\phi}}{\vdash \psi \rightarrow [negotiation; correction]\Box\phi}$$

The left branch proves that if ϕ holds during *negotiation*, an open condition $Lv_0 + C_0 < m$ should be satisfied. The proof is shown as follows:

$$\begin{array}{ll} & \psi \vdash Lv_0 + C_0 < m \\ \hline \text{P10} & \psi \vdash \forall t \geq 0 (t \leq L \rightarrow v_0 t + C_0 < m) \\ \hline \text{D3} & \psi \vdash \forall t \geq 0 \{\!\!\{getDisplacement\text{🔔}, m - currentdis < ST, \\ & \qquad C = v_0 t + C_0 \wedge l = t, true\}\!\!\}\phi \\ \hline \text{D7} & \psi \vdash [getDisplacement\text{🔔}, m - currentdis < ST, true, \\ & \qquad C_{time} = v_0 \wedge l_{time} = 1]\phi \\ \hline \text{T3} & \psi \vdash [negotiation]\Box\phi \end{array}$$

In the proof above, in the step applying P10, L is replaced by $\forall t \geq 0$ to obtain a general case.In the step applying D3, $v_0 t + C_0$ is substituted by C and t is substituted by l. In the step applying D7, $C = v_0 t + C_0$ and $l = t$ are special solutions of differential equations $C_{time} = v_0$ and $l_{time} = 1$ respectively.

The right branch proves that if ϕ continues to holds after *negotiation* has completed when continuing with an adjusted acceleration a, an open condition

$v_0^2 < 2b(m - Lv_0 - C_0) \wedge Lv_0 + C_0 < m$ should be satisfied. The proof is shown as follows:

$$\text{P10}\ \frac{\psi, t \geq 0 \vdash v_0^2 < 2b(m - Lv_0 - C_0) \wedge Lv_0 + C_0 < m}{\psi, t \geq 0 \vdash \{\!|getDisplacement\text{🔔}, m - currentdis < ST, C = v_0 t + C_0 \wedge l = t, true|\!\}}$$

$$\text{D3}\ \frac{\forall \tilde{t} \geq 0 (l \leq L \rightarrow -\frac{b}{2}\tilde{t}^2 + v_0\tilde{t} + C_0 < m)}{\psi, t \geq 0 \vdash \{\!|getDisplacement\text{🔔}, m - currentdis < ST, C = v_0 t + C_0 \wedge l = t, true|\!\}}$$

$$\text{T3,D7}\ \frac{\forall \tilde{t} \geq 0 \{\!|MAControl\text{🔔}, m - currentdis < ST, MAControl\boxtimes \wedge C = -\frac{b}{2}\tilde{t}^2 + v_0\tilde{t} + C_0, true|\!\}\phi}{\psi, t \geq 0 \vdash \{\!|getDisplacement\text{🔔}, m - currentdis < ST, C = v_0 t + C_0 \wedge l = t, true|\!\}}$$

$$\text{P3}\ \frac{[MAControl\text{🔔}, m - currentdis < ST, MAControl\boxtimes, C_{time} = V \wedge V_{time} = -b]\Box\phi}{\psi \vdash t \geq 0 \rightarrow \{\!|getDisplacement\text{🔔}, m - currentdis < ST, C = v_0 t + C_0 \wedge l = t, true|\!\}}$$

$$\text{P10}\ \frac{[correction]\Box\phi}{\psi \vdash \forall t \geq 0 \{\!|getDisplacement\text{🔔}, m - currentdis < ST, C = v_0 t + C_0 \wedge l = t, true|\!\}}$$

$$\text{D7}\ \frac{[correction]\Box\phi}{\psi \vdash [negotiation][correction]\Box\phi}$$

In the proof above, in the first step applying P10, $\forall \tilde{t} \geq 0$ is substituted to obtain a general case. In the step applying D3, $-\frac{b}{2}\tilde{t}^2 = v_0\tilde{t} + C_0$ is substituted by C. In the step applying T3 and D7, $-\frac{b}{2}\tilde{t}^2 = v_0\tilde{t} + C_0$ is a special solution of differential equation set $C_{time} = V, V_{time} = -b$.

5 Concluding Remarks and Related Work

In this paper, we extended SRML semantic domain by defining HL^2TSs and it's hybrid traces which represent the system evolution. Based on this, we made a formal extension of SRML, which includes the extension of the language of business role and the language of business protocol. For our case study, we specified a Train-Control system with SRML and verified a safety constraint of it.

This work has been done mainly on the basis of [6,12]. In [6] a formal specification of SRML is provided. In [12], hybrid programs and the logic dTL for reasoning about the temporal behaviour of hybrid programs are defined, and a set of calculus for deductive verification is provided. In the SRML extension, we redefined hybrid programs and dTL formulas over the extended SRML semantic domain, thus enables the evolution of a service-oriented transition system with continuous traces can be reasoned about with the calculus in [12]. Furthermore, to define the HL^2TSs, we referenced [13] for Hybrid Automata; to redefine hybrid programs and dTL, we referenced [14,15] for Temporal Logic and Dynamic Logic basis. Different from various approaches for modeling and verifying hybrid systems, such as that provided in [17,18], our approach deals with hybrid transition systems, which integrate interactions among components with hybrid systems.

Although our work extends SRML, which is defined to specify service-oriented transition systems, it does not include the content of service discovery and binding. In [16] a formal operational semantics for service discovery and binding is brought forward. A prospect of our future work might be applying continuous time execution to this approach.

Appendix 1: A Rule Schema of the dTL Calculus

A rule schema of the dTL calculus can be found in Table 1. In these rules, $\phi, \psi \in Fml$ and $\pi \in Fml_T$; $\chi_1 \in EF$ and $\chi_2 \in E\text{-}EF$, $T_1, \ldots, T_n \in ETERM$, $T \in E\text{-}ETERM$ and $t \in ETERM_{time}$. In D3, M is a first-order formula and the substitution of $M^{T'_1 \ldots T'_n}_{T_1 \ldots T_n}$, which replaces $T_1 \ldots T_n$ by $T'_1 \ldots T'_n$ in M. In D7-D8,

Table 1. Rule schemata of the temporal dynamic dTL verification calculus

$$(P1)\ \frac{\vdash \phi}{\neg\phi \vdash} \qquad (P2)\ \frac{\phi \vdash}{\vdash \neg\phi}$$

$$(P3)\ \frac{\phi \vdash \psi}{\vdash \phi \rightarrow \psi} \qquad (P4)\ \frac{\phi, \psi \vdash}{\phi \wedge \psi \vdash}$$

$$(P5)\ \frac{\vdash \phi \quad \vdash \psi}{\vdash \phi \wedge \psi} \qquad (P6)\ \frac{\vdash \phi \quad \psi \vdash}{\phi \rightarrow \psi \vdash}$$

$$(P7)\ \frac{\phi \vdash \quad \psi \vdash}{\phi \vee \psi \vdash} \qquad (P8)\ \frac{\vdash \phi, \psi}{\vdash \phi \vee \psi}$$

$$(P9)\ \frac{}{\phi \vdash \phi} \qquad (P10)\ \frac{M_0 \vdash G_0}{M \vdash G}$$

$$(D1)\ \frac{[\beta]\pi \wedge [\gamma]\pi}{[\beta \cup \gamma]\pi} \qquad (D2)\ \frac{\langle\beta\rangle\pi \vee \langle\gamma\rangle\pi}{\langle\beta \cup \gamma\rangle\pi}$$

$$(D3)\ \frac{M^{T'_1 \ldots T'_n}_{T_1 \ldots T_n}}{\{trigger, guard, T_1 = T'_1 \wedge \ldots \wedge T_n = T'_n \wedge \chi_1, \chi_2\}M} \qquad (D4)\ \frac{\{\beta\}\{\gamma\}\phi}{\{\beta;\gamma\}\phi}$$

$$(D5)\ \frac{\phi \wedge [\beta;\beta^*]\phi}{[\beta^*]\phi} \qquad (D6)\ \frac{\phi \vee \langle\beta;\beta^*\rangle\phi}{\langle\beta^*\rangle\phi}$$

$$(D7)\ \frac{\forall t \geq 0[trigger, guard, v = f_{v_0}(T_1, \ldots, T_n, t) \wedge \chi_1, \chi_2]\phi}{[trigger, guard, \chi_1, v_{time} = T \wedge \chi_2]\phi}$$

$$(D8)\ \frac{\exists t \geq 0\langle trigger, guard, v = f_{v_0}(T_1, \ldots, T_n, t) \wedge \chi_1, \chi_2\rangle\phi}{\langle trigger, guard, \chi_1, v_{time} = T \wedge \chi_2\rangle\phi}$$

$$(T1)\ \frac{[\beta]\Box\phi \wedge [\beta][\gamma]\Box\phi}{[\beta;\gamma]\Box\phi} \qquad (T2)\ \frac{\langle\beta\rangle\Diamond\phi \vee \langle\beta\rangle\langle\gamma\rangle\Diamond\phi}{\langle\beta;\gamma\rangle\Diamond\phi}$$

$$(T3)\ \frac{[\beta]\phi \wedge [first\beta]\phi}{[\beta]\Box\phi} \qquad (T4)\ \frac{\langle\beta\rangle\phi}{\langle\beta\rangle\Diamond\phi}$$

$$(T5)\ \frac{[\beta;\beta^*]\Box\phi}{[\beta^*]\Box\phi} \qquad (T6)\ \frac{\langle\beta;\beta^*\rangle\Diamond\phi}{\langle\beta^*\rangle\Diamond\phi}$$

f_{v_0} is the solution of the initial value problem $v_{time} = T, v^{\triangle(\sigma(0))} = v_0$, where $\sigma(0)$ is the state in which variable v has the value v_0. In P10, $Cl_{\forall}(F_0 \rightarrow G_0) \rightarrow Cl_{\forall}(F \rightarrow G)$ is an instance of a first-order tautology of real arithmetic and $Cl_{\forall}$ is the universal closure.

Appendix 2: SRML Specification of Module *Train-Control*

The SRML specification of service module *Train-Control* is shown in Fig. 6.

MODULE TRAINCONTROL **is**

DATATYPES

sorts: speed, acceleration, displacement, time

PROVIDES

RBC: RadioBlockCentre

REQUIRES

BA: Balise

COMPONENTS

TR: Train

WIRES

TR Train		**TB**		**BA** Balise
rcv getDisplacement 🔒dis	R i₁	Straight. I(displacement)	S i₁	**snd** reportDisplacement 🔒 dis

RBC RadioBlockCentre		**RT**		**TR** Train
rcv getPosition 🔒 p	R i₁	Straight. I(displacement)	S i₁	**snd** reportPosition 🔒 p
s&r systemBusy	S	Straight	R	**r&s** MAControl
s&r newMA	S	Straight	R	**r&s** moveOn

END MODULE

Fig. 6. Module *Train-Control*

References

1. Georgakopoulos, D., Papazoglou, M.: Service-Oriented Computing. The MIT Press Cambridge, Massachusetts (2009)
2. van der Schaft, A., Schumacher, H.: An Introduction to Hybrid Dynamical Systems. Springer, London, UK (1999)
3. Abreu, J., Bocchi, L., Fiadeiro, J.L., Lopes, A.: Specifying and composing interaction protocols for service-oriented system modelling. In: Derrick, J., Vain, J. (eds.) FORTE 2007. LNCS, vol. 4574, pp. 358–373. Springer, Heidelberg (2007)
4. Building Systems using a Service Oriented Architecture. White paper version 0.9 (2005)
5. Fiadeiro, J., Lopes, A., Bocchi, L., Abreu, J.: The SENSORIA reference modelling language. In: Wirsing, M., Hölzl, M. (eds.) SENSORIA. LNCS, vol. 6582, pp. 61–114. Springer, Heidelberg (2011)
6. Abreu, J.: Modelling business conversations in service component architectures. Ph.D thesis, University of Leicester (2009)
7. Yu, N.: Injecting continuous time execution into service-oriented computing. Ph.D thesis, Ludwig-Maximilians-Universität München (to be appeared)
8. Platzer, A.: Towards a hybrid dynamic logic for hybrid dynamic systems. In: Blackburn, P., Bolander, T., Braüner, T., de Paiva, V., Villadsen, J. (eds.) LICS International Workshop on Hybrid Logic 2006, Seattle USA, ENTCS (2007)
9. Henzinger, T., Nicollin, X., Sifakis, J., Yovine, S.: Symbolic model checking for read-time systems. In: LICS, pp. 394–406. IEEE Computer Society (1992)
10. Alonso, G., Casati, F., Kuno, H., Machiraju, V.: Web Services. Springer, New York (2004)
11. ter Beek, M.H., Fantechi, A., Gnesi, S., Mazzanti, F.: An action/state-based model-checking approach for the analysis of communication protocols for service-oriented applications. In: Leue, S., Merino, P. (eds.) FMICS 2007. LNCS, vol. 4916, pp. 133–148. Springer, Heidelberg (2008)
12. Platzer, A.: AVACS - Automatic verification and analysis of complex systems. Technical report No. 12, AVACS (2007)
13. Henzinger, T.A.: The theory of hybrid automata. In: Lnan, M.K., et al. (eds.) LICS 1996, vol. 170, pp. 256–292. Springer, Berlin (2000)
14. Emerson, E.A.: Temporal and modal logic. In: van Leeuwen, J. (ed.) HTCS, vol. A, pp. 995–1072. Elsevier, Amsterdam (1995)
15. Harel, D., Kozen, D.: Dynamic Logic. The MIT Press Cambridge, Massachusetts (2000)
16. Fiadeiro, J., Lopes, A., Bocchi, L.: An abstract model of service discovery and binding. In: Formal Aspects of Computing, vol. **23**, pp. 433–463. Springer, Berlin (2011)
17. Fadlisyah, M., Ölveczky, P.C., Ábrahám, E.: Formal modeling and analysis of human body exposure to extreme heat in HI-maude. In: Durán, F. (ed.) WRLA 2012. LNCS, vol. 7571, pp. 139–161. Springer, Heidelberg (2012)
18. Quesel, J., Mitsch, S., Loos, S., Aréchiga, N., Platzer, A.: How to model and prove hybrid systems with KeYmaera: A tutorial on safety. In: STTT (2015)
19. European Train Control System (ETCS) Open Proofs - Open Source. http://openetcs.org/

Author Index

Printed in the United States
By Bookmasters